U0068771

可靠度工程概論

楊善國　編著

 全華圖書股份有限公司

國家圖書館出版品預行編目資料

可靠度工程概論 / 楊善國編著. -- 六版. – 新北市:
　全華圖書, 2019.12
　　面 ; 　公分
　ISBN 978-986-503-310-1(精裝)

　1.數理統計

319.5　　　　　　　　　　　　108020832

可靠度工程概論

作者 / 楊善國

發行人 / 陳本源

執行編輯 / 林昱先

出版者 / 全華圖書股份有限公司

郵政帳號 / 0100836-1 號

印刷者 / 宏懋打字印刷股份有限公司

圖書編號 / 0571275

六版一刷 / 2019 年 12 月

定價 / 新台幣 400 元

ISBN / 978-986-503-310-1 (精裝)

全華圖書 / www.chwa.com.tw

全華網路書店 Open Tech / www.opentech.com.tw

若您對書籍內容、排版印刷有任何問題，歡迎來信指導 book@chwa.com.tw

臺北總公司(北區營業處)
地址：23671 新北市土城區忠義路 21 號
電話：(02) 2262-5666
傳真：(02) 6637-3695、6637-3696

中區營業處
地址：40256 臺中市南區樹義一巷 26 號
電話：(04) 2261-8485
傳真：(04) 3600-9806

南區營業處
地址：80769 高雄市三民區應安街 12 號
電話：(07) 381-1377
傳真：(07) 862-5562

初版作者序

　　民國八十五年開始，作者即在本校開設「可靠度工程」的選修課。當時可靠度的課程在國內還不熱門，近年來則逐漸受到重視。在這些年的教學過程中，坊間可靠度方面的原文書有一些，但中文書卻一直少見。因而作者將這些年教學的材料編輯成冊，希望有助於國內這方面的教學與研究所需。

　　可靠度的領域包羅甚廣，本書本文的內容僅為粗淺的概論。較深入的主題如失效分析、預防保養、裴氏網路、以及失效預測等的探討則放在附錄裡。附錄 A～E 是本人過去所發表的幾篇論文，讀者可視需要選讀，或再由文後的參考文獻去找尋相關資料深入研讀。

　　感謝前太空計劃室主任、現任建國科技大學徐佳銘校長以及中華民國品質學會可靠度委員會主任委員張起明博士的撥冗題序。作者才疏學淺，文中恐有謬誤，祈請先進賢達不吝指正，謝謝。願上帝祝福您！

楊善國　謹誌

於國立勤益技術學院機械系

初版序一

PREFACE

　　可靠度的觀念源自第二次世界大戰中軍品使用時之有效性問題，那是性命交關的大事。戰後與性命相關的用品如飛機、升降機、纜車、核能電廠，無法保養維護的器械如人造衛星、太空船，會嚴重影響生產效率的工業機器人、彈性製造系統，甚至交通工具及一般民生用品等，也相繼採用可靠度工程來確保在預定的使用環境及時間下安全無虞地發生效用，除保證壽命之外，更避免發生意外事故、臨時故障及降低維護工作與壽命週期成本。

　　由於現代的工業產品及系統功能越來越多，對軟硬體功能及性能的整合性要求越來越高，使系統複雜度也越來越高，未經利用可靠性工程的設計製造很少能讓系統發揮其應有的效用，而更由於系統的研發時間越來越短，如何使研發人員充分瞭解及應用可靠度工程的確需要一本好書，而本書就是其中之一。

　　本書內容涵蓋可靠度工程的基本理論─機率與統計，再介紹可靠度與失效分析，評估與試驗，並就可能的提高可靠度方法加以整理陳述，可使讀者容易地獲得產品或系統不可或缺的可靠度工程實用技術，非常難得。

　　鑑於可靠度工程已逐漸由軍用品擴大到廣泛的工業產品與民生用品，而且由少數大系統逐漸擴大到眾多的小系統，其重要性與被重視程度越來越高，已經成為工程科系學生及專業人士不可不學的專門知識，本書的出版正符此需要，特在此予以推薦。

建國科技大學　校長

徐佳銘　謹誌

2004.12.16

初版序二

可靠度工程技術源自於國防工業，由於航太及武器系統之設計精密，性能要求高，造價較昂貴，一旦發生失效，所造成的損失必極為嚴重，因此國防工業界鑑於產品之安全保障與信譽維護，特發展出可靠度工程技術加以應用，其後商用產品業界亦逐漸體認其重要性而廣泛加以學習應用，以促使產品不發生故障與避免失效。綜觀世界各國可靠度工程技術之發展應用歷史，美歐日如此，我國亦不例外，國內中山科學研究院擔負國防武器系統研發，其可靠度工程技術研究與應用，已具相當基礎與深度；並與國內相關學(協)會共同推動可靠度教育與訓練，國內大專院校近年來，因有多位教授致力於可靠度工程技術之教學與研究，培育了許多具有可靠度技術專長之人才，對國內工業產品可靠度之提升深具貢獻。

本書作者楊善國博士敬虔事天、為人謙和、學有專精。他曾在中山科學研究院任 IDF 戰機空用儀電系統研究多年，隨後在國立勤益科技大學任教，並於 1999 年在國立交通大學完成博士學位，他的博士論文與可靠度與失效分析之研究有關，他對裴氏網路經(Petri-net approach)研究有獨到之見解，發表過多篇國際期刊論文。他以在國內完整的學經歷，將他的研究成果投注在學術發表、教學論壇以及在對國內產業界之推廣應用上，表現卓越，令人欽佩。

有鑑於國內產業升級正處在關鍵性階段，產業界對改善產品品質與可靠性的技術需求甚殷，本人服務於中山科學研究院，有機會參與可靠度工作；涉獵可靠度工程技術，並兼任國立交通大學機械工程研究所之可靠度工程教學及在中華民國品質學會擔任可靠度委員會主任委員，此三合一之角色促使本人對推動國內可靠度與維護度等工程技術善盡棉薄心力。欣聞本書作者楊博士近期已

V

獲教育部教授升等，並受邀於印度召開之 2005 國際可靠度與安全度等工程會議；擔任論文審查委員及大會演溝等，現今又將他長期之研究心得融會於他的新書－可靠度工程概論，本書內容具實用性且參考價值高，將可提供國內大專院校可靠度工程教學教材，以造就國內大專院校之可靠度技術專長人才。本人甚覺與有榮焉並樂予為序。

沈起明

2005.04.20

四版序

　　初識作者楊善國教授是在 1999 年受邀擔任他博士論文口試委員，那時才赫然發覺可靠度工程也可與自動化科技結合，而為相當有意義的研究，著實擴展了我的眼光。自那時起，楊教授常謙然稱我為師，讓我受之有愧；其實他見多識廣，瞭解的東西比我還多。經多年交往，我逐漸發覺他是國內少數自大學開始即學習自動化工程的學者，曾受過幾位國內非常知名自動化與控制學者的薰陶。因學有專精，楊教授曾於中山科學研究院等國防部所屬單位任職與任教過，其餘大部份時間則在大專院校作育英才，並從事自動化量測與可靠度工程相關研究，研教成果相當豐碩，近年來常受邀至國際研討會與中國大陸各大學演講或授課。

　　楊教授於教學、研究工作外，也曾擔任系、院、校級行政主管，忙碌之餘，還抽空撰寫此書，實屬難能可貴。本書應為楊教授在自動控制領域外，涉獵可靠度工程教學、研究與實務二十多年來的經驗記載，其內容包括可靠度概論、機率概要、統計概要、可靠度與失效分析、可靠度評估、可靠度試驗等章節，書後並以附錄形式呈現多篇可靠度工程實際應用案例及相關機率與統計圖表。個人以為初學者可以透過此書瞭解可靠度工程所需的數學及其可能探討的範疇，而複學或複習者則可藉研讀此書擴展他們對可靠度工程瞭解的層面，至少我個人曾獲得後者效果。

　　楊教授告訴我，此書因蒙讀者青睞，將再版印刷，託我致序，個人感到相當榮幸，並樂撰此序就教楊教授與讀者們。

國立臺灣大學機械工程學系/工業工程學研究所

吳文方 謹致

五版序

PREFACE

　　可靠度的觀念起源可追溯到第二次世界大戰，美國陸軍從武器的維修耗時及壽期維護成本不扉意識到可靠度的需要，另由於武器是軍人的第二生命，為了要滿足戰場的嚴格要求，於是乃自 1940 年起針對提昇武器的可靠度進行了很多的研究工作。可靠度的技術發展於 1950 年代，美國國防部為解決戰場之極低的裝備開箱存活率與過高的維修備份零件需求，針對電子裝備特別成立一個可靠度特別調查委員會(Advisory Group on the Reliability of Electronic Equipment, AGREE)，處理軍品可靠度問題。歷經多年的發展美軍訂立完整的可靠度工作定義與架構，例如系統與裝備之發展與生產可靠度計畫(MIL-STD-785B)作為軍用標準之依據規範足見可靠度工程在軍事上的重要性。時至今日，重視產品可靠度需求程度高漲，可靠度工程技術也已從軍品擴大到飛機、船鑑、鐵路、捷運等大眾交通運輸工具系統以及講求高可靠度的產品與重要的生產設施上的廣泛運用。

　　本書作者楊善國教授在本校任教多年，為人謙和、治學態度嚴謹，平時主要著墨在可靠度工程技術等專業的研究領域上，並多年協助國內產業界從事提昇可靠度的技術推廣與應用上貢獻卓著且成效斐然，足為可靠度研究領域的學術界表率。此次欣聞楊博士再版可靠度工程概論，將多年的研究成果融入書中，以深入淺出的方式編撰，不論工程科系的在學學子或是在職的或專業人士都是一本非常適用的專業書籍，由於本書再版非常符合學界與業界的期待，特別在此予以榮譽推薦。

國立勤益科技大學特聘教授

管理學院創院長

中華訓練品質學會理事長

林文燦 博士

編輯部序

PREFACE

　　「系統編輯」是我們的編輯方針，我們所提供給您的，絕不只是一本書，而是關於這門學問的所有知識，它們由淺入深，循序漸進。

　　因國防工業之航太及武器系統的設計精密、要求高性能、造價昂貴等，若失效造成的損失相當嚴重，進而發展出可靠度工程技術。本書以追求系統整合及經濟效益的角度來闡述物料管理，作者將長期的研究心得融會於書中，內容在如何提昇產品品質，用以降低生產成本並提高生產力，相當具實用性且參考價值高，可提供作國內大專院校可靠度工程教材，以造就可靠度技術之專長人才。

目 錄

Contents

第 1 章 概 論

第 2 章 機率概要

第 3 章　統計概要

第 4 章　可靠度與失效分析

第 5 章　可靠度評估

第 6 章　可靠度試驗

附錄 A

附錄 B

附錄 C

Chapter 1

概　論

■ 1.1　源起

可靠度的研究肇始於二次世界大戰期間納粹德國對於 V-1 火箭的開發。而美軍運往亞洲戰區的軍火有許多在運送途中或儲存期間就已經失效,在運抵戰區時無法發揮功效,以致嚴重影響戰力。當時因沒有可靠度的觀念,誤認為問題是製造時作業員的疏忽或品檢員的檢驗不力所致。在美國電子設備可靠度顧問團(Advisory Group on Reliability of Electronic Equipment, AGREE)提出該問題的研究報告之後,才發現事實並非如此,其實一產品的品質與時間、操作環境、要求效能等都有密切關係,因而也正式開始了可靠度研究的新頁。

■ 1.2　概念

人們注重產品的品質,並且採取種種管制手段以維持和改進品質。然而僅只依靠品質管制並不足以保證產品在現場的操作條件下能夠長期發揮其預定

機能。因為產品的原始設計可能不足以達成使產品在不同的現場操作條件下長期不失效運轉的目標。原因之一是設計者在產品的設計階段沒有將操作條件或環境壓力或操作壽命等因素納入考量，也就是沒有「可靠度(Reliability)」的觀念。

可靠度是產品在「既定的時段內」、「要求的環境狀況下」能順利發揮其「預設機能」的「機率」。

可靠度理論如同統計品管，是以機率與統計為基礎的學域，其內涵大致可歸納為三：「可靠度數學」、「可靠度工程」、與「可靠度管理」。可靠度數學是描述可靠度的工具和語言；可靠度工程是執行可靠度設計、製造、測試、與驗證等的專業技術；而將各部門有效率地結合，使可靠度方案能圓滿達成目標，則為可靠度管理的領域。

■ 1.3　定義與對象

可靠度亦稱為「機能品管(Functional quality control)」或「時間維度的品質(Time-dimensional quality)」。其定義為：「產品在預定時段或任務時間內、於要求之壓力環境下可發揮其足夠績效的條件機率」，至少具有下列要素：

1.　以機率(Probability)表示。

2.　可以衡量及測試，以評估該產品的績效(Performance)。

3.　涉及預測產品的預測壽命(Lifetime)。

4.　必須界定產品的操作環境條件(Environment Stress)。

例如：「本產品有 95% 的機率以大於 90% 的產能在溫度(70±10)°F 及相對溼度 60% 的無塵環境下不失效地運作 100 小時」。

此處所謂產品可區分為下列不同層次：

1.　系統(System)：泛指所有執行任務所需的一切軟、硬體，包括主要設備及周邊設備。

2. 子系統(Subsystem)或稱裝備(Equipment)：為系統的一部份，具有執行系統中一部分功能的能力。

3. 模組(Module)或稱裝置(Device)：為子系統的一部份，在子系統中具有單獨特定功能。

4. 組件(Component)：為模組的一部份，由基本的零件組合而成，是系統中最低階的組合體。

5. 零件(Part)：是系統中最基本的單元，無法再與以分解。

所以各層次之間的關係為：系統 ⊃ 次系統 ⊃ 模組 ⊃ 組件 ⊃ 零件。可靠度問題的探討可針對每一層次進行之。

■ 1.4　可靠度日趨重要的原因

1. 商品耐久化
2. 商品複雜化
3. 自動化
4. 工程規模巨大化
5. 研發的時效限制

概念 → 設計 → 原型(Proto type) → 測試 → 除錯 → 測試 → 量產 → 銷售 → 使用者 → 丟在路邊也沒人要撿 ≈ 12 個月

■ 1.5　壽命週期成本(Life-Cycle Cost, LCC)

選用一產品時之成本考量應以壽命週期成本為準，而不應僅以售價之購入成本為唯一考量。如表 1-1 之例，某產品於招標時三家供應商之售價不同，若僅以售價考量則甲供應商可得標，然若考慮由購入開始至產品壽命終結所支出之總額，則應選擇丙供應商，此即「壽命週期成本」。

表 1-1　壽命週期成本

支出項目	金額(單位：千元)		
	供應商 甲	供應商 乙	供應商 丙
售　價	42	60	47
人　工	129	116	84
備　品	40	30	30
文書作業	12	18	12
能　源	255	225	245
管　理	60	45	50
訓　練	10	6	8
停　機	80	100	70
壽命週期成本	628	600	546

❖ 習 題

1. 請敘述「可靠度」之定義，並列舉說明其內涵要素。

2. 可靠度的內涵分為哪三大領域？

3. 請闡述「壽命週期成本的觀念」。

機率概要

■ 2.1 緒言

1. 可靠度之定義係指「一產品在某既定時段及環境下，以要求的產能且不失效運轉的機率」，故必須藉機率為工具來討論可靠度的問題。

2. 一產品為可靠狀態表示其不為失效狀態，可靠與失效彼此互斥。而一產品之失效係隨機出現，故失效現象只能用機率模式表示。

3. 一事件依其隨機與否之特性可分成下列兩類：

   (1) Deterministic(決定)：指該事件可以數學式分析描述者，又可分為
      a. Periodic(週期性)：依一定規則週而復始出現者；
      b. Aperiodic(非週期性)：無一定反覆週期，或說週期為無窮大者。

   (2) Stochastic(推測)或 Random(隨機)：無法以數學式描述規範者，又可分為

a. Ergodic(代表性)：在有限時間內對該事件觀測所得之資料即可代表全體資料者；

b. Non-ergodic(非代表性)

隨機事件通常以機率統計的方法對其分析處理。

■ 2.2　機率概念

設 X 表示一事件，則其發生的機率表示為：P(X)，P(X)有下列特性：

1.　$P(X) = \lim\limits_{N \to \infty} \dfrac{n}{N}$

N：測試總數

n：在 N 次測試中，X 事件發生的次數。

2.　$0 \le P(X) \le 1$。

3.　若 $\bar{X}$ 表示 X 事件的反面，例如 X 為失效，$\bar{X}$ 即為不失效，則 $P(\bar{X}) = 1 - P(X)$。

4.　符號「∩」代表交集、及、AND，則

(1)　$P(X \cap Y)$ 代表 X 事件及 Y 事件同時發生的機率

(2)　$P(X \cap Y) = P(Y \cap X)$

5.　符號「∪」代表聯集、或、OR，則

(1)　$P(X \cup Y)$：X 事件或 Y 事件其中之一發生的機率

(2)　$P(X \cup Y) = P(Y \cup X)$

6.　$P(X \mid Y)$ 表示在 Y 事件發生的前提下，X 事件會發生的機率。

∵ $P(X \cap Y) = P(X \mid Y)P(Y)$，∴ $P(X \mid Y) = \dfrac{P(X \cap Y)}{P(Y)}$，且 $P(X \cap Y) = P(Y \mid X)P(X)$。

7.　若 X 與 Y 為彼此相依(Mutually dependent)，則

(1)　$P(X \cap Y) = P(X \mid Y)P(Y) = P(Y \mid X)P(X)$

(2)　$P(X \cup Y) = P(X) + P(Y) - P(X \cap Y)$

邏輯關係可由如圖 2-1 之范氏圖(Venn diagram)看出。

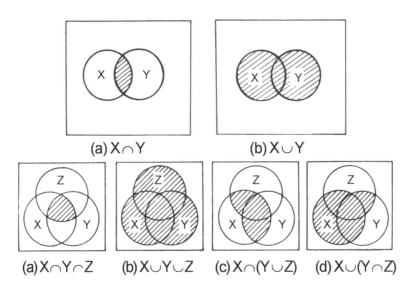

圖 2-1　二事件與三事件的范氏圖

8.　若 X 與 Y 為彼此獨立(Mutually independent)，則

(1)　$P(X \cap Y) = P(X)P(Y)$

(2)　$P(X \mid Y) = P(X)$

(3)　$P(Y \mid X) = P(Y)$

(4)　$P(X \cup Y) = P(X) + P(Y) - P(X)P(Y) = 1 - P(\overline{X})P(\overline{Y})$

9.　若 X 與 Y 為彼此互斥(Mutually exclusive)，則

(1)　$P(X \mid Y) = P(Y \mid X) = 0$

(2)　$P(X \cap Y) = 0$

(3)　$P(X \cup Y) = P(X) + P(Y)$

例 2.1　某斷路器的需求時未開的失效(Failure to open on demand)機率爲 0.02。現將兩個該型斷路器串聯：

1. 若失效爲獨立，求：
   (1) 該系統失效的機率？
   (2) 至少一個斷路器失效的機率？
2. 若已知在第一個斷路器失效的狀況下，第二個也失效的機率爲 0.1，求：
   (1) 該系統失效的機率？
   (2) 至少一個斷路器失效的機率？

解　設 X 爲第一個斷路器失效的事件，Y 爲第二個斷路器失效的事件。

已知：$P(X) = P(Y) = 0.02$

1. X，Y 爲獨立
   (1) $P(X \cap Y) = P(X)P(Y) = 0.02 \times 0.02 = 0.0004$
   (2) $P(X \cup Y) = P(X) + P(Y) - P(X)P(Y) = 0.02 + 0.02 - (0.02 \times 0.02)$
       $= 0.0396 = 1 - [1 - P(X)][1 - P(Y)] = 1 - (1 - 0.02)(1 - 0.02)$
       $= 0.0396$
2. X，Y 爲相依：$P(Y \mid X) = 0.1$
   (1) $P(X \cap Y) = P(Y \mid X)P(X) = 0.1 \times 0.02 = 0.002$
   (2) $P(X \cup Y) = P(X) + P(Y) - P(Y \mid X)P(X) = 0.02 + 0.02 - 0.1 \times 0.02$
       $= 0.038$

---

10. $P(X \cup Y \cup Z) = P(X) + P(Y) + P(Z) - P(X \cap Y) - P(X \cap Z) - P(Y \cap Z) + P(X \cap Y \cap Z)$

11. 若 X、Y、Z 爲彼此獨立，則

    $P(X \cup Y \cup Z) = P(X) + P(Y) + P(Z) - P(X)P(Y) - P(X)P(Z) - P(Y)P(Z) + P(X)P(Y)$
    　　　　　　$P(Z)$，

或

$$P(X \cup Y \cup Z) = 1 - [P(\overline{X})P(\overline{Y})P(\overline{Z})] = 1 - [1 - P(X)][1 - P(Y)][1 - P(Z)]$$

12. 若 X_1、X_2、……、X_n 為彼此獨立，則

(1) $P(X_1 \cap X_2 \cap \cdots \cap X_n) = P(X_1)P(X_2)\cdots P(X_n) = \displaystyle\prod_{i=1}^{n} P(X_i)$

(2) $P(X_1 \cup X_2 \cup \cdots \cup X_n) = 1 - [P(\overline{X_1})P(\overline{X_2})\cdots P(\overline{X_n})]$

$= 1 - [1 - P(X_1)][1 - P(X_2)]\cdots[1 - P(X_n)] = 1 - \displaystyle\prod_{i=1}^{n} \left[1 - P(X_i)\right]$

例 2.2　　一機翼之接縫需釘入 28 枚相同的鉸釘以接合，若其中有任一鉸釘施作不良則需全部重作。根據過去紀錄，有 18% 的接縫重作。問：

1. 若缺點為獨立，則任一鉸釘施作不良之機率為多少？

2. 若欲將重作率降至 5% 以下，則每一枚鉸釘之施作不良率應為多少？

解　　設 X_i 表示第 i 枚鉸釘施作不良，且已知 $P(X_1) = P(X_2) = \cdots = P(X_{28})$

1. $P(X_1 \cup X_2 \cup \cdots \cup X_{28}) = 1 - [1 - P(X_1)]^{28} = 0.18$，

    $\Rightarrow$　$P(X_1) = 0.0071 = P(X_2) = \cdots = P(X_{28})$

2. $1 - [1 - P(X_1)]^{28} = 0.05$，

    $\Rightarrow$　$P(X_1) = 0.0018 = P(X_2) = \cdots = P(X_{28})$

---

13. 事件組合的邏輯運算關係如表 2-1 所示。

表 2-1　邏輯運算表

編號	運算	說明
1	$X \cap Y = Y \cap X$	交換律
2	$X \cup Y = Y \cup X$	
3	$X \cap (Y \cap Z) = (X \cap Y) \cap Z$	結合律
4	$X \cup (Y \cup Z) = (X \cup Y) \cup Z$	
5	$X \cap (Y \cup Z) = (X \cap Y) \cup (X \cap Z)$	分配律
6	$X \cup (Y \cap Z) = (X \cup Y) \cap (X \cup Z)$	
7	$X \cap (X \cup Y) = X$	吸收律
8	$X \cup (X \cap Y) = X$	
9	$X \cap X = X$	
10	$X \cup X = X$	
11	$X \cap \bar{X} = \phi$ (ϕ：空集合)	互餘
12	$X \cup \bar{X} = I$ (I：全集合)	
13	$\overline{(\bar{X})} = X$	
14	$X \cap \phi = \phi$	
15	$X \cup \phi = X$	
16	$X \cap I = X$	
17	$X \cup I = I$	
18	$\overline{(X \cap Y)} = \bar{X} \cup \bar{Y}$	狄摩根定律
19	$\overline{(X \cup Y)} = \bar{X} \cap \bar{Y}$	
20	$X \cup (\bar{X} \cap Y) = X \cup Y$	
21	$\bar{X} \cap (X \cup \bar{Y}) = (\bar{X} \cap \bar{Y}) = \overline{(X \cup Y)}$	

■ 2.3 獨立失效(Independent failure)

1. 需求(Demand)：要求產品發揮其預期功能，例：燈泡點亮、引擎啓動、繼電器作動……，有下列特性：

(1)　需求的次數須能計數或可推算估計。

(2)　每次需求的成功機率須與過去已需求的次數為獨立關係。

2.　設 R_n 為系統在 n 次需求後仍能發揮功能之機率(i.e.可靠度)，X_n 為事件 X 在第 n 次需求為成功事件，則 $R_n = \mathrm{P}(X_1)\mathrm{P}(X_2)\cdots\cdots\mathrm{P}(X_n)$。

又若 $\mathrm{P}(X_1) = \mathrm{P}(X_2) = \cdots\cdots = \mathrm{P}(X_n) = q$，i.e. $\mathrm{P}(\overline{X_n}) = 1 - q = p$ (失效機率)，則

$R_n = q^n = (1-p)^n \quad \Rightarrow \quad \ln R_n = n\ln(1-p)$

若 $p \ll 1$，則 $\ln(1-p) \approx -p$，　$\therefore\ \ln R_n = -np \quad \Rightarrow \quad R_n = e^{-np}$

由此可知，可靠度 R_n 隨需求次數 n 成指數遞減。

3.　設 Δt 為每兩次需求間之平均時間間隔，則需求 n 次之總遞減時間

$t = n\Delta t \quad \Rightarrow \quad n = t/\Delta t$

$\Rightarrow \quad R_n(t) = e^{-\frac{t}{\Delta t}p} = e^{-\frac{p}{\Delta t}t} = e^{-\lambda t}$，$\lambda = p/\Delta t$ 稱為失效率(Failure rate)。

例 2.3　某閥門製造商知道其產品在操作 100 次之後有 1%的閥門會失效，若失效為隨機且獨立，求：

1.　每次需求的失效機率？

2.　設計壽命為 500 次需求的可靠度？

3.　若每週使用該閥三次，則該閥在第一年內會失效的機率？

解　1.　100 次操作之後的成功率(可靠度)

$= 1 - 1\% = 0.99 = R_{100} = e^{-100p} \quad \Rightarrow \quad p \approx 10^{-4}$

2.　$R_{500} = e^{-500p} = \exp(-500 \times 10^{-4}) = 0.95$

3.　操作的平均時間間隔 $= 1/(3 \times 52) = 0.0064$(年)

$\lambda = 10^{-4}/0.0064 = 0.015625 \quad \Rightarrow \quad R(t) = \exp(-0.015625t)$

所以該閥在第一年內會失效的機率

$= 1 - R(1) = 1 - \exp(-0.015625 \times 1) = 0.0155$

---

4. 設某系統爲 m 個組件的非複聯(Non-redundant)設計,且其中任一組件失效則整個系統失效。令 $P(X_m)$ 爲第 m 個組件成功運作的機率,且 $P(X_m) = q$,其失效機率 $P(\overline{X_m}) = 1 - q = p$,則該系統之可靠度 $R_m = P(X_1)P(X_2) \cdots$ $P(X_m) = (1 - p)^m \Rightarrow R_m = e^{-mp}$,所以可靠度亦隨系統複雜度(組件個數 m)增加而呈指數遞減。

例 2.4 某電路板含有 100 個二極體,該電路板製造商知道該電路板有 5% 因獨立的二極體失效而未能通過品管,因爲任一個二極體失效將導致該電路板失效。求:

1. 任一個二極體的失效機率?

2. 若電路板擴大爲含 500 個該型二極體,則該電路板未能通過品管的百分比爲多少?

3. 若欲將 500 個二極體的電路板的失效率也保持爲 5%,則任一個二極體的失效機率應降低爲多少?

解

1. $R_{100} = e^{-100p} = 1 - 5\% = 0.95 \Rightarrow p = 0.5 \times 10^{-3}$

2. $R_{500} = e^{-500p} = \exp(-500 \times 0.5 \times 10^{-3}) = 0.7788 \Rightarrow$ 未能通過品管的百分比 $= 1 - 0.7788 = 22.12\%$

3. $R_{500} = e^{-500p0} = 0.95 \Rightarrow p0 = 0.1 \times 10^{-3}$

■ 2.4 隨機變數 ─ 離散型 (Discrete,或稱不連續 Discontinuous)

1. 出象(Outcome):一行爲(如實驗、測試……等)的結果。

2. 隨機試驗(Random experiment):無法預知出象的試驗。

   (1) 其出象可分:

   　　a. 定性描述:如產品爲良品或不良品。

　　　b. 定量描述：如良品或不良品的個數。

(2)　其所有可能出象所成之集合稱「樣本空間(Sample space, S)」。

3.　隨機變數(Random variable)

(1)　S 中之任一元素 S_i 經一函數 X 賦予一實數 $X(S_i)$，則 $X(S_i)$ 即稱爲隨機變數。

(2)　若 $X(S_i)$ 所對應之值爲離散數值，如 1, 3, 8……(通常爲整數)，則該類隨機變數稱爲「離散隨機變數(或計數值變數)」。

4.　令 $X(S_i)= X_i$，則每一 X_i 均對應一出現的機率 $f(X_i)$，函數 f 稱爲隨機變數 X_i 之「機率分配函數(Probability distribution function，pdf)」簡稱「機率函數(Probability function)」，對離散隨機變數而言則稱「機率質量函數(Probability mass function)」，且 $\sum_i f(X_i) = 1$。

5.　累積分配函數(Cumulative distribution function)：$F(X_i)$
定義：$F(X_i) = \sum_{j=0}^{i} f(X_j)$，即隨機變數出現爲小於或等於 X_i 之機率的和，

i.e. $F(X_i) = P(X \le X_i)$。

6.　期望值(Expected value，或稱平均值 Mean value, Average)：μ 或 $E(X)$
定義：$\mu = \sum_i X_i f(X_i) = E(X)$

其意義爲：所有變數各乘以其對應出現機率所得之積的總和；即在出現機率的加權考慮下變數的平均值。

例 2.5　　求一枚均勻骰子出現點數的期望值。

解　　因每一面出現的機率均爲 1/6，所以
$$\mu = (1 \times 1/6) + (2 \times 1/6) + (3 \times 1/6) + (4 \times 1/6) + (5 \times 1/6) + (6 \times 1/6)$$
$$= 3.5$$

---

7. 變異數(Variance)：σ^2 或 $V(X)$

    定義：$\sigma^2 = \sum\limits_{i} (X_i - \mu)^2 f(X_i) = V(X)$

    $\sigma^2 = V(X) = E[(X - \mu)^2] = E[X^2] - \mu^2$

    證明：

    $$\sigma^2 = \sum\limits_{i} (X_i - \mu)^2 f(X_i) = E\left[\left(X_i - \mu\right)^2\right]$$

    $$= \sum\limits_{i} (X_i^2 - 2X_i\mu + \mu^2) f(X_i)$$

    $$= \sum\limits_{i} X_i^2 f(X_i) - 2\mu \sum\limits_{i} X_i f(X_i) + \mu^2 \sum\limits_{i} f(X_i)$$

    $$= E[X^2] - 2\mu \times \mu + \mu^2$$

    $$= E[X^2] - \mu^2$$

8. 標準差(Standard deviation)：σ

    其意義爲：以μ爲中心，變數分布的範圍。

9. 離散機率分配

    (1) 二項分配(Binominal distribution)

        a. 變數僅分成發生與不發生兩種的二分法。

        b. 設發生的機率爲 p，不發生的機率爲 q，則 $p + q = 1$。

        c. 機率函數 $f(X) = \dfrac{n\,!}{X\,!(n-X)\,!}\, p^X q^{n-X}$ ，

           其中 n：事件進行的總次數(Number of trail)

           X：要求事件發生的次數(Number of demand)，

           $X = 0, 1, 2, \cdots\cdots, n$。

           Note：$\dfrac{n\,!}{X\,!(n-X)\,!} = C_X^n = \binom{n}{X}$

d. $\mu = np$

證明：

$\mu = E(X) = \sum_i X_i f(X_i)$

$= \sum_i X_i \begin{pmatrix} n \\ X_i \end{pmatrix} p^{X_i} q^{n-X_i} = \sum_i X_i \frac{n\,!}{(n-X_i)\,!\,X_i\,!} p^{X_i} q^{n-X_i}$

$= np \sum_i \frac{(n-1)\,!}{(n-X_i)\,!\,(X_i-1)\,!} p^{X_i-1} q^{n-X_i}$

$= np \sum_i \begin{pmatrix} n-1 \\ X_i-1 \end{pmatrix} p^{X_i-1} q^{(n-1)-(X_i-1)}$

$= np \sum_i f(Y_i)$　　　(Note：$Y_i = X_i - 1$)

$= np$

e. $\sigma^2 = npq$

證明：

$\sigma^2 = V(X) = E[(X-\mu)^2] = E[X^2] - \mu^2$

$= E[X(X-1) + X] - (np)^2$

$= E[X(X-1)] + E[X] - (np)^2$

$= E[X(X-1)] + np - (np)^2$

$= \sum_i X_i(X_i-1) f(X_i) + np - (np)^2$

$= \sum_i X_i(X_i-1) \frac{n\,!}{(n-X_i)\,!\,X_i\,!} p^{X_i} q^{n-X_i} + np - (np)^2$

$= n(n-1) p^2 \sum_i \frac{(n-2)\,!}{(n-X_i)\,!\,(X_i-2)\,!} p^{X_i-2} q^{n-X_i} + np - (np)^2$

$= n^2 p^2 - np^2 + np - (np)^2$

$= np(1-p)$

$= npq$

例 2.6　某型飛機有四個輪胎，根據記錄平均每 1200 次著陸有一次爆胎事件。若爆胎事件為獨立，且至多二輪胎爆胎均可安全著陸。求：

1. 此型飛機因爆胎事件未能安全著陸的機率。
2. 該分配的期望值及變異數。

解　設 X 為爆胎數，爆胎機率為 p，輪胎個數為 n，i.e. $p = 1/1200 = 0.00083$, $n = 4$, $q = 1 - p = 0.99917$

1. 飛機安全著陸的機率為

   (0 個爆胎的機率 + 1 個爆胎的機率 + 2 個爆胎的機率)

   $$= f(0) + f(1) + f(2) = F(2) = P(X \le 2)$$

   $$= \frac{4!}{4!} \left(0.00083 \right)^0 \left(0.99917 \right)^4 + \frac{4!}{3!} \left(0.00083 \right)^1 \left(0.99917 \right)^3$$

   $$+ \frac{4!}{2!\,2!} \left(0.00083 \right)^2 \left(0.99917 \right)^2 = 0.9999999977$$

   未能安全著陸的機率 $= 1 - 0.9999999977 = 2.3 \times 10^{-9}$

2. $\mu = np = 4 \times 0.00083 = 0.00332$，

   $\sigma^2 = npq = 4 \times 0.00083 \times 0.99917 = 0.003317$

例 2.7　已知某型壓縮幾之失效機率 p 為 0.1，現有 10 架該型壓縮機受測，求：

1. 失效架數的期望值，
2. 失效架數的變異數，
3. 完全無失效的機率，
4. 兩架以上失效的機率。

解　設 X 為失效架數

1. $E(X) = \mu = np = 10 \times 0.1 = 1$
2. $\sigma^2 = npq = np(1 - p) = 10 \times 0.1 \times 0.9 = 0.9$

3. $P(X=0) = \begin{pmatrix} 10 \\ 0 \end{pmatrix} p^0 (1-p)^{10} = 1 \times 1 \times (1-0.1)^{10} = 0.349$

4. $P(X \geq 2) = 1 - f(0) - f(1) = 1 - 0.349 - [10 \times 0.1 \times (1-0.1)^9] = 0.2636$

(2) 波瓦松分配(Poisson distribution)

　　a. 適用於失效機率非常小，受測品個數非常多的場合。

　　b. 機率函數 $f(X) = \dfrac{m^X e^{-m}}{X!}$ ， $\mu = m$ ， $\sigma^2 = m$

例 2.8　某產品故障機率 $p = 0.001$，求 2000 件受測品中測得 3 件故障的機率。

(1) 以二項分配： $f(3) = \dfrac{2000!}{(2000-3)! \, 3!} (0.001)^3 (0.999)^{1997} = 0.1805$

(2) 以波瓦松分配： $\mu = n \times p = 2000 \times 0.001 = 2$ ，

　　 $f(3) = \dfrac{2^3 e^{-2}}{3!} = 0.1804$

例 2.9　以波瓦松近似值求解例 2.7。

(1) $E(X) = \mu = np = 10 \times 0.1 = 1$

(2) $\sigma^2 = \mu = 1$

(3) $P(X = 0) = e^{-m} = e^{-1} = 0.3678$

(4) $P(X \geq 2) = 1 - f(0) - f(1) = 1 - 2e^{-m} = 0.2642$

■ 2.5 隨機變數－連續型 (Continuous，或稱類比 Analog)

1. 連續型隨機變數又稱計量值變數。

2. 相關函數

    (1) 機率密度函數(Probability density function)：$f(x)$

        a. If $\Delta x \approx 0$, $P(x \leq X \leq x + \Delta x) = f(x)\,\Delta x$

        b. $\int_{a}^{b} f(x)dx = P(a \leq X \leq b)$

    (2) 累積分配函數(Cumulative distribution function)：$F(x)$

        a. $F(x) = P(X \leq x) = \int_{-\infty}^{x} f(y)dy$ or $f(x) = \dfrac{d}{dx}F(x)$

        b. $\int_{-\infty}^{\infty} f(x)dx = P(-\infty \leq X \leq \infty) = 1$

        c. $\because F(x) = 0,\ \forall t < 0,\quad \therefore F(t) = \int_{0}^{t} f(x)dx,\quad \int_{0}^{\infty} f(t)dt = 1$

    (3) 期望值：$E\big[g(x)\big] = \int_{-\infty}^{\infty} g(x)f(x)dx$

3. 連續機率分配函數

    (1) 常態分配(Normal distribution)

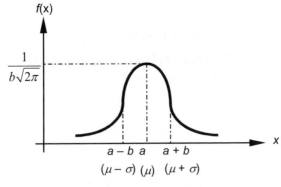

圖 2-2　常態分配曲線

a. 機率密度函數：$f(x) = \dfrac{1}{b\sqrt{2\pi}} \exp\left[-\dfrac{1}{2}\left(\dfrac{x-a}{b}\right)^2 \right], \quad -\infty < x < \infty$

　　通常以 $N(a, b^2)$ 表示，i.e. $f(x) = N(a, b^2)$

b. 累積分配函數

$$F(x) = \int_{-\infty}^{x} f(y)dy = \int_{-\infty}^{x} \dfrac{1}{b\sqrt{2\pi}} \exp\left[-\dfrac{1}{2}\left(\dfrac{y-a}{b}\right)^2 \right]dy$$

c. $f(x)$ 的曲線呈鐘形分配

d. $f(x)$ 左右對稱於 $x = a$,　i.e. $f(a+t) = (a-t)$

e. $f(a) = \dfrac{1}{b\sqrt{2\pi}}$ 爲極大值

f. $f(x)$ 在 $x = a+b$ 及 $x = a-b$ 處爲反曲點(Point of inflection)

g. $f(\pm\infty) \rightarrow 0$, i.e.　$f(x)$ 以 x 軸爲漸進線

h. $\mu = a, \sigma^2 = b^2$

i. 以 μ 爲中心的面積：$A(\mu \pm 1\sigma) = 68.26\% \times A(\mu \pm \infty)$

$$A(\mu \pm 2\sigma) = 95.46\% \times A(\mu \pm \infty)$$

$$A(\mu \pm 3\sigma) = 99.73\% \times A(\mu \pm \infty)$$

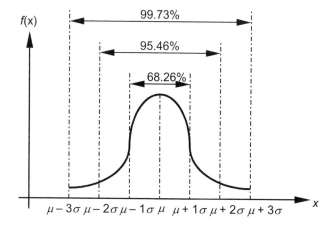

圖 2-3　常態分配以 μ 爲中心不同 σ 所佔的面積

(2) 標準常態分配(Normalized normal distribution)

  a. 令 $Z = \dfrac{x-\mu}{\sigma}$ 則

    (a) 若 $x = \mu$ $\Rightarrow$ $Z = \dfrac{x-\mu}{\sigma} = 0$

    (b) 若 $x = \sigma$ $\Rightarrow$ $Z = \dfrac{\sigma-0}{\sigma} = 1$

    可將 $N(\mu, \sigma^2)$ 標準化(Normalization)成 $N(0, 1)$，

    即 $\mu = 0$，$\sigma = 1$

  b. 現 $F(x) = \displaystyle\int_{-\infty}^{x} f(y)dy = \int_{-\infty}^{x} \dfrac{1}{\sigma\sqrt{2\pi}} \exp\left[-\dfrac{1}{2}\left(\dfrac{y-\mu}{\sigma}\right)^2 \right]dy$，

    令 $Z = \dfrac{y-\mu}{\sigma}$ 則

    $\dfrac{dZ}{dy} = \dfrac{1}{\sigma}$ $\Rightarrow$ $dZ = \dfrac{dy}{\sigma}$ $\Rightarrow$

    $F(Z) = \displaystyle\int_{-\infty}^{z} \dfrac{1}{\sqrt{2\pi}} \exp\left[-\dfrac{1}{2}Z^2 \right]dZ$

    $= \Phi\left(\dfrac{y-\mu}{\sigma}\right) = \Phi(Z)$

    $\Rightarrow$ $f(Z) = \dfrac{1}{\sqrt{2\pi}} \exp\left[-\dfrac{1}{2}Z^2 \right]$

  c. $\because f(Z)$ 左右對稱，$\therefore f(Z) = f(-Z)$ 且 $F(-Z) = 1 - F(-Z)$

  d. 中央極限定理(Central Limit Theorem, CLT)

    當一個變數受到許多相加的變化來源影響時，無論這些變化的分配情形如何，相加後的分配係趨近常態分配。

  e. 常態分配符合重疊定理(Law of superposition)

    設 $f(x) = N(\mu_x, \sigma_x^2)$，$f(y) = N(\mu_y, \sigma_y^2)$，若 $z = ax + by$

    $(a, b$ 為常數)，則

    $f(z) = N(\mu_z, \sigma_z^2) = N(a\mu_x + b\mu_y, a^2\sigma_x^2 + b^2\sigma_y^2)$

例 2.10 某機器的有用壽命可用常態分配 $N(500, 100^2)$ 描述。求：

1. 該型機器可不失效地操作至少 600 小時的機率？
2. 已知已不失效地操作 500 小時，還能再操作至少 100 小時而不失效的機率？

解 1. $P(T > 600) = P\left(\dfrac{T-500}{100} > \dfrac{600-500}{100}\right) = P(Z > 1) = 1 - 0.8413$

$= 0.1587$

2. $P(T > 600 \mid T > 500) = \dfrac{P\left(T > 600 \cap T > 500\right)}{P\left(T > 500\right)} = \dfrac{P\left(T > 600\right)}{P\left(T > 500\right)}$

$= \dfrac{P\left(Z > \dfrac{600-500}{100}\right)}{P\left(Z > \dfrac{500-500}{100}\right)} = \dfrac{P\left(Z > 1\right)}{P\left(Z > 0\right)} = \dfrac{0.1587}{0.5} = 0.3174$

例 2.11 設某切割工具的刀刃因磨耗而損壞的時間可用 $N(2.8, 0.6^2)$ 描述，單位為小時。求：

1. 該刀刃在低於 1.5 小時內磨損的機率？
2. 多久更換刀具一次可使刀刃失效機率低於 10%？

解 1. $P(T < 1.5) = P\left(\dfrac{T-2.8}{0.6} < \dfrac{1.5-2.8}{0.6}\right) = P(Z < -2.1667)$

$= \Phi(-2.1667) = 0.0151 = 1.51\%$

2. $P(T < t) = 0.1 = P\left(\dfrac{T-2.8}{0.6} < \dfrac{t-2.8}{0.6}\right) = P\left(Z < \dfrac{t-2.8}{0.6}\right) = \Phi\left(\dfrac{t-2.8}{0.6}\right)$

查表知 $\dfrac{t-2.8}{0.6} \cong -1.28 \quad \Rightarrow \quad t = 2.03$ 小時

(3) 對數常態分配(Lognormal distribution)

    a. 適用於隨機變數由數個隨機變數相乘積而成的場合，
例：$X = x_1 \times x_2 \times \cdots \times x_n$

    b. $\ln X = \ln\left(x_1 \times x_2 \times \cdots \times x_n\right) = \ln x_1 + \ln x_2 + \cdots + \ln x_n$，
根據中央極限定理，若 n 夠大則 $\ln X$ 應為常態分配。

    c. 常態分配之累積分配函數為：$F\left(y\right) = \int_0^y \frac{1}{\sigma\sqrt{2\pi}} \exp\left[-\frac{1}{2}\left(\frac{y-\mu}{\sigma}\right)^2\right] dy$

現令該函數中之隨機變數 $y = \ln x$，則 $\frac{dy}{dx} = \frac{1}{x}$ $\Rightarrow$ $dy = \frac{dx}{x}$

$\Rightarrow$ 該函數變為：$F\left(x\right) = \int_0^y \frac{1}{\sigma x\sqrt{2\pi}} \exp\left[-\frac{1}{2}\left(\frac{\ln x-\mu}{\sigma}\right)^2\right] dx$

可得對數常態分配之機率密度函數

$$f(x) = \frac{1}{\sigma x\sqrt{2\pi}} \exp\left[-\frac{1}{2}\left(\frac{\ln x-\mu}{\sigma}\right)^2\right]$$ ；

亦即 y 呈常態分配，則 x 呈對數常態分配。

    d. $\sigma_{LN}^{\;2} = \ln\left[\left(\frac{\sigma}{\mu}\right)^2 + 1\right]$，$\mu_{LN} = \ln\mu - \frac{1}{2}\sigma_{LN}^{\;2}$

$\Rightarrow$ $\mu = \exp\left(\mu_{LN} + \frac{1}{2}\sigma_{LN}^{\;2}\right)$，

$\sigma^2 = \exp\left(2\mu_{LN} + 2\sigma_{LN}^{\;2}\right) - \exp\left(2\mu_{LN} + \sigma_{LN}^{\;2}\right)$

推導：

$\mu_{LN} = \ln\mu - \frac{1}{2}\sigma_{LN}^{\;2}$ $\Rightarrow$ $\ln\mu = \mu_{LN} + \frac{1}{2}\sigma_{LN}^{\;2}$

$\Rightarrow$ $\mu = \exp\left(\mu_{LN} + \frac{1}{2}\sigma_{LN}^{\;2}\right)$……(1)

$\sigma_{LN}^{\;2} = \ln\left[\left(\frac{\sigma}{\mu}\right)^2 + 1\right]$ $\Rightarrow$ $\exp\left(\sigma_{LN}^{\;2}\right) = \left(\frac{\sigma}{\mu}\right)^2 + 1$

$$\Rightarrow \quad \exp\left(\sigma_{LN}{}^2\right) - 1 = \frac{\sigma^2}{\mu^2}$$

$$\Rightarrow \quad \sigma^2 = \mu^2 \times \exp\left(\sigma_{LN}{}^2\right) - \mu^2 \cdots\cdots(2)$$

(1)代入(2)得

$$\sigma^2 = \exp\left(2\mu_{LN} + \sigma_{LN}{}^2\right) \times \exp\left(\sigma_{LN}{}^2\right) - \exp\left(2\mu_{LN} + \sigma_{LN}{}^2\right)$$

$$= \exp\left(2\mu_{LN} + 2\sigma_{LN}{}^2\right) - \exp\left(2\mu_{LN} + \sigma_{LN}{}^2\right)$$

例 2.12 測試 100 輛汽車之剎車壽命，得其平均失效里程爲 56669.5 哩，標準差爲 12393.64 哩。

1. 設其爲 Normal distribution，求：其壽命低於 50000 哩之機率？
2. 若其爲 Lognormal distribution，求：其壽命低於 50000 哩之機率？

解

1. $Z = \dfrac{x - \mu}{\sigma} = \dfrac{x - 56669.5}{12393.64}$，

$$P(x < 50000) = P\left(Z < \frac{50000 - 56669.5}{12393.64}\right) = P(Z < -0.5381) = 0.2950$$

2. $\sigma_{LN}{}^2 = \ln\left[\left(\dfrac{\sigma}{\mu}\right)^2 + 1\right] = \ln\left[\left(\dfrac{12393.64}{56669.5}\right)^2 + 1\right] = 0.0467$

$$\Rightarrow \quad \sigma_{LN} = 0.2162$$

$$\mu_{LN} = \ln\mu - \frac{1}{2}\sigma_{LN}{}^2 = \ln 56669.5 - \frac{1}{2}\,0.0467$$

$$= 10.945 - 0.0234 = 10.9216$$

$$P(x < 50000) = P\left(Z < \frac{\ln x - \mu_{LN}}{\sigma_{LN}}\right)$$

$$= P\left(Z < \frac{\ln 50000 - 10.9216}{0.2162}\right)$$

$$= P(Z < -0.4710) = 0.3190$$

❖ 習 題

1. 經測試 100 輛汽車之剎車壽命，得平均失效里程爲 56669.5 哩，標準差爲 12393.64 哩。

    (1) 設其爲 Normal distribution，求此剎車壽命低於 50000 哩之機率？

    (2) 若改爲 Lognormal distribution，求此剎車壽命低於 50000 哩之機率？

2. 經測試 100 輛汽車之剎車壽命，得平均失效里程爲 50000 哩，標準差爲 10000 哩。

    (1) 設其爲 Normal distribution，求此剎車壽命低於 40000 哩之機率？

    (2) 若改爲 Lognormal distribution，求此剎車壽命低於 40000 哩之機率？

3. 某型飛機有四具彼此獨立且完全相同之引擎，設每具引擎之可靠度爲 0.9。若至少有兩具引擎正常工作則飛機可成功飛行，求就引擎而言該型飛機之可靠度？

4. 設某型飛彈之可靠度爲 0.85：

    (1) 若每次發射間爲彼此獨立，則同時發射兩枚而其中任一枚擊中目標之機率爲若干？

    (2) 若每次發射間爲相依，且若第一枚未擊中則第二枚擊不中之機率爲 0.2；若第一枚擊中則第二枚亦可擊中之機率爲 0.85，求同時發射兩枚至少一枚擊中之機率？

5. 某產品故障機率 $p = 0.005$，求 1000 件中有 3 件故障的機率。

    (1) 其故障爲二項分配，

    (2) 其故障爲 Poisson 分配。

6. 設某隨機試驗成功的機率爲 p，失敗機率爲 q，且 $q = 1-p$。現共進行 n 次試驗，請證明二項分配的(1)期望值 $\mu = np$，(2)變異數 $\sigma^2 = npq$。

7. 圖中 a，b，c，d 四個開關，其中任一開關可成功閉合之機率為 0.8，且彼此獨立。求：

   (1) X，Y 間可構成通路的機率，

   (2) 若已知 X，Y 間有通路，求 a 與 b 均可成功閉合的機率。

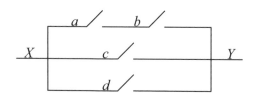

8. 某工作母機之刀具壽命可以 N(2.8, (0.6)^2)描述，其中單位為小時。求：

   (1) 刀具壽命長於 3 小時的機率，

   (2) 若欲使加工時遇見刀具失效而停機的機率在 0.2%以下，保養人員應隔多久就更換刀具一次？

Chapter 3

統計概要

■ 3.1 緒言

對於一未知的事物(特別是 Stochastic 事物)，通常藉下列流程對其進行分析，以期窺其全貌。

事物 —實驗測試→ 數據 —統計→ 機率函數 —機率→ 分析

圖 3-1 事物分析流程

經對該事物進行測試(或實驗)而取得數據之後，須由數據判斷可描述該事物的統計分配。由已知數據推論該數據為何種分配，將面對三種不確定性(Uncertainty)：

1. 固有不確定性(Intrinsic uncertainty)：例如汽車耗油量會因車輛及駕駛人而異，此變異並非來自量測技術或儀器之精準度。反之，量測某些基本物理

量時，每次所得數據有所不同係由量度的重現性與隨機誤差所造成，非固有不確定性。

2. 模式的選擇(Model selection)：該以何種機率模式來描述該已知數據？

3. 參數的決定(Parameter determination)：參數乃指機率分配的特徵值，例如期望值μ及變異數σ^2；另表示樣本分配的特徵值，如樣本平均數$\bar{x}$及樣本變異數 S^2，稱為統計量。而統計量是隨機樣本的函數，故對不同的隨機樣本將得到不同的統計值。

■ 3.2 模式選擇

1. 機率分配的選用有下列方法供選擇：

(1) 引用其他類似的事物，在過去經證實為正確的分配。

(2) 依現象的特性來決定，例：

　　a. 現象為多個效應的和，則適用 Normal distribution。

　　b. 現象為多個效應的積，則適用 Lognormal distribution。

(3) 繪統計圖表：

　　a. 失效統計表，如表 3-1。

表 3-1　某產品之失效統計表

失效區間(小時)		堪用數		區間失效統計		累計失效數	累計失效數百分比	累計倖存數百分比
始	終	始	終	數量	百分比(%)			
100	110	25	23	2	8	2	8	92
110	120	23	19	4	16	6	24	76
120	130	19	7	12	48	18	72	28
130	140	7	1	6	24	24	96	4
140	150	1	0	1	4	25	100	0
總計				25	100			

b. 直方圖(Bar chart) ，如圖 3-2。

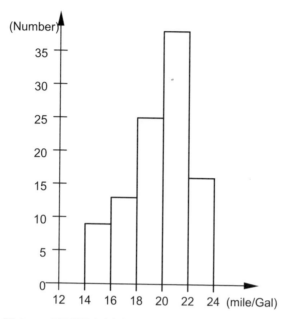

圖 3-2 某型汽車耗油量測試結果之直方圖(組數 5)

c. 失效數量百分比圖，如圖 3-3。

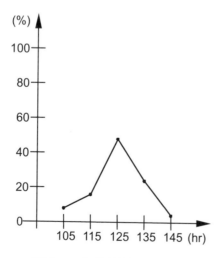

圖 3-3 失效數量百分比圖

d.　累計失效數量百分比圖，如圖 3-4。

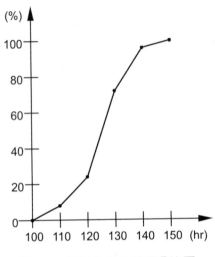

圖 3-4　累計失效數量百分比圖

e.　累計倖存數量百分比圖，如圖 3-5。

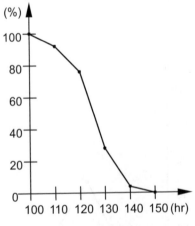

圖 3-5　累計倖存數量百分比圖

2.　例子

100 輛某型汽車耗油量數據如表 3-2，請根據該表的數據繪製直方圖。

表 3-2　某型汽車耗油量測試數據(單位：哩/加侖)

19.0	21.5	20.2	21.6	20.2	22.6	17.3	21.7	14.5	18.7	22.6	20.5	*14.1*	19.2	19.3	22.7	22.9	22.8	21.7	20.4
15.1	18.8	22.1	17.4	21.6	15.1	17.5	19.7	19.0	16.5	22.9	19.2	18.3	20.5	21.7	16.4	18.6	19.1	18.7	21.9
18.2	14.3	22.8	21.9	20.5	19.2	22.7	21.7	15.3	19.7	21.9	19.2	16.2	17.1	22.5	*23.9*	16.3	21.4	17.1	19.6
19.1	20.9	15.5	22.4	21.8	18.5	19.3	20.4	21.1	20.2	15.1	21.6	21.5	18.2	20.6	21.3	21.1	20.2	16.8	22.4
21.9	20.5	18.9	20.3	20.8	18.1	16.6	*23.9*	22.0	20.7	16.8	18.8	22.7	21.3	20.8	21.1	20.0	16.9	14.7	21.9

繪製直方圖的步驟如下：

(1) 找出數據之最大值 H 與最小值 L：本例之 H=23.9，L=14.1。

(2) 求數據之展幅(Span)S：S=H－L=23.9－14.1=9.8

(3) 決定分組數目：數據分成太多或太少組均不易看出數據分配的情形，組數的決定有下列二法可供參考：

a. 史特吉思(Sturges)所提出之經驗公式：

$$k = 1 + 3.32 \log(n) \quad\text{.. (3.1)}$$

其中 $k =$ 分組數，$n =$ 數據數目。

b. 經驗值表，如表 3-3。

表 3-3　分組數之經驗表

$n =$ 數據數目	50 – 100	100 – 250	250 以上
$k =$ 分組數	6 – 10	7 – 14	10 – 20

(4) 決定組界：最小組界應比數據之最小值略小，而組界之值應比數據之小數點少一位，以免發生數據剛好落於組界上而不知該屬於上或下一組的情形。

(5) 計算每組所屬數據之個數並據此繪出直方圖。

本例分成 5 組之結果示於圖 3-2，分成 10 組與 20 組的結果則分別如圖 3-6 與 3-7 所示。

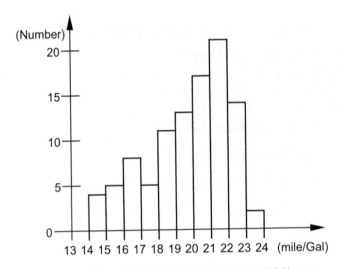

圖 3-6　某型汽車耗油量測試結果之直方圖(組數 10)

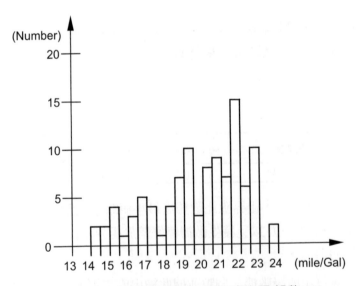

圖 3-7　某型汽車耗油量測試結果之直方圖(組數 20)

　　繪製直方圖的目的之一為藉直方圖之外形(Envelope)判別其為何種分配，但這項工作現在可交由許多市面可購得之套裝軟體來執行。

■ 3.3　參數估計

1. 估計類型

   (1)　點估計(Point estimation)：求最佳估計值

   (2)　區間估計(Interval estimation)：求信任區間

2. 點估計：利用樣本數據以估計得到一數值作爲參考值

   (1)　平均數μ的估計

   a. 設 $x_1, x_2, ..., x_n$,爲得自分配 $f(x)$ 的 n 個隨機樣本，則 $\bar{x} = \dfrac{1}{n}\sum_{i=1}^{n} x_i$ 稱爲 $f(x)$ 的平均數μ的估計量(Estimator)。(The statistics, calculated from the samples that are used to estimate population parameters, are called *estimators*.)若可得 m 個 $\bar{x}$，則任一 $\bar{x}_1$，$\bar{x}_2$，...，$\bar{x}_m$ 稱爲 $f(x)$ 的平均數μ的估計值(Estimate)。

   b. 一個好的 Estimator 須有下列性質：

      (a) *Unbiased*(不偏的)：The estimator $\hat{\theta}$ is an unbiased estimator for a parameter θ if and only if $E(\hat{\theta}) = \theta$. In other words, an unbiased estimator should not consistently underestimate nor overestimate the true value of the parameter.

      (b) *Consistent*(一致的)：A consistent estimator is one that is unbiased and converges more closely to the true value of the population parameter as the sample size increases. In other words, the estimator $\hat{\theta}$ is said to be a consistent estimator of θ if the probability of making errors of any given size ε, tends to zero as n (sample size) tends to infinity — that is, $P\left[\left|\hat{\theta} - \theta\right| > \varepsilon\right] \to 0 \; as \; n \to \infty$, for any fixed positive ε.

(c) *Efficient*(效率的)：An efficient estimator is a consistent estimator whose standard deviation is smaller than the standard deviation of any other estimator for the same population parameter. Efficiency is measured by

Relative efficiency = $V\left(\hat{\theta}_2\right)/V\left(\hat{\theta}_1\right)$, where $V\left(\hat{\theta}_1\right)$ and $V\left(\hat{\theta}_2\right)$ are the variance of the two estimators $\hat{\theta}_1$ and $\hat{\theta}_2$ for the same population parameter: the better estimator has the smaller variance; and

The asymptotic relative efficiency (ARE): the relative efficiency is a function of n and to avoid this the asymptotic relative efficiency is adopted as $ARE = \underset{n\to\infty}{Lim}\dfrac{V\left(\hat{\theta}_2\right)}{V\left(\hat{\theta}_1\right)}$.

(d) *Sufficient*(足夠的)：A sufficient estimator is an estimator that utilizes all the information about the parameter that the sample possesses.

c. 判斷估計量優劣的兩個機率理論：

(a) 不論 $f(x)$ 爲何種分配，只要 $f(x)$ 的變異數 σ^2 存在，則估計量 $\bar{x}$ 的變異數 $\sigma_{\bar{x}}^2$ 等於 $\dfrac{\sigma^2}{n}$ (i.e. $\sigma_{\bar{x}}^2 = \dfrac{\sigma^2}{n}$)。據此可知，若樣本量 n 越大則估計誤差越小。

(b) 不論 $f(x)$ 爲何種分配，只要樣本量 n 夠大則 $\bar{x}$ 的抽樣分配將成常態分配，此即中央極限定理。原應爲：$N\left(\mu_{\bar{x}}, \sigma_{\bar{x}}^2\right)$，根據經驗若 $n \geq 30$ 則 $\bar{x}$ 即是 $N\left(\mu, \dfrac{\sigma^2}{n}\right)$。

(2) 標準差(S)的估計

$S_{\mu}^2 = \dfrac{1}{n}\sum_{i=1}^{n}(x_i - \mu)^2$，若 μ 爲已知，則 S_{μ}^2 爲不偏估計量(Unbiased estimator)，i.e. $E(S_\mu^2) = \sigma^2$。但 μ 及 σ^2 通常均為未知，故以 $\overline{x}$ 取代 μ，則

$S_{\overline{x}}^2 = \dfrac{1}{n}\displaystyle\sum_{i=1}^{n}(x_i - \overline{x})^2$ 可能不再有不偏性，i.e. $E(S_{\overline{x}}^2) \neq \sigma^2$。此時須改寫成

$S_{\overline{x}}^2 = \dfrac{1}{n-1}\displaystyle\sum_{i=1}^{n}(x_i - \overline{x})^2$ (自由度少 1)，則 $E(S_{\overline{x}}^2) = \sigma^2$。

證明：

$$\overline{x} = \frac{1}{n}\sum_{i=1}^{n} x_i \implies \sum_{i=1}^{n} x_i = n\overline{x} \cdots\cdots(1)$$

$$\sigma^2 = \frac{1}{n}\sum_{i=1}^{n}(x_i - \mu)^2 \implies \sum_{i=1}^{n}(x_i - \mu)^2 = n\sigma^2 \cdots\cdots(2)$$

$$\sum_{i=1}^{n}(x_i - \mu) = (x_1 - \mu) + (x_2 - \mu) + ... + (x_n - \mu)$$

$$= (x_1 + x_2 + ... + x_n) - n\mu$$

$$= \sum_{i=1}^{n} x_i - n\mu = n\overline{x} - n\mu \text{ (根據(1)式)}$$

$$= n(\overline{x} - \mu) \cdots\cdots(3)$$

$$E\left[\sum_{i=1}^{n}(x_i - \mu)(\overline{x} - \mu)\right] = E\left[(\overline{x} - \mu)\sum_{i=1}^{n}(x_i - \mu)\right] = E\left[(\overline{x} - \mu)n(\overline{x} - \mu)\right] \text{ (根據(3)式)}$$

$$= n(\overline{x} - \mu)^2 \cdots\cdots(4)$$

$$= (\overline{x} - \mu)^2 = \left(\frac{1}{n}\sum_{i=1}^{n} x_i - \mu\right)^2 = \left(\frac{1}{n}\sum_{i=1}^{n} x_i - \frac{n\mu}{n}\right)^2 = \left[\frac{1}{n}\sum_{i=1}^{n}(x_i - \mu)\right]$$

$$= \frac{1}{n^2}\sum_{i=1}^{n}(x_i - \mu)^2 = \frac{1}{n^2}(n\sigma^2) \text{ (根據(2)式)}$$

$$= \frac{1}{n}\sigma^2 \cdots\cdots(5)$$

$$E\left(S_{\overline{x}}^2\right) = E\left(\frac{1}{n-1}\sum_{i=1}^{n}(x_i - \overline{x})^2\right)$$

$$= \frac{1}{n-1}E\left\{\sum_{i=1}^{n}\left[(x_i - \mu) - (\overline{x} - \mu)\right]^2\right\}$$

$$= \frac{1}{n-1}E\left\{\sum_{i=1}^{n}\left[(x_i - \mu)^2 - 2(x_i - \mu)(\overline{x} - \mu) + (\overline{x} - \mu)^2\right]\right\}$$

$$= \frac{1}{n-1} \left\{ E\left[\sum_{i=1}^{n}(x_i-\mu)^2\right] - 2E\left[\sum_{i=1}^{n}(x_i-\mu)(\overline{x}-\mu)\right] + E\left[\sum_{i=1}^{n}(\overline{x}-\mu)^2\right] \right\}$$

$$= \frac{1}{n-1} \left\{ n\sigma^2 - 2E\left[\sum_{i=1}^{n}(x_i-\mu)(\overline{x}-\mu)\right] + E\left[\sum_{i=1}^{n}(\overline{x}-\mu)^2\right] \right\} \quad \text{(根據(2)式)}$$

$$= \frac{1}{n-1} \left\{ n\sigma^2 - 2n(\overline{x}-\mu)^2 + E\left[\sum_{i=1}^{n}(\overline{x}-\mu)^2\right] \right\} \quad \text{(根據(4)式)}$$

$$= \frac{1}{n-1} \left\{ n\sigma^2 - 2n(\overline{x}-\mu)^2 + n(\overline{x}-\mu)^2 \right\}$$

$$= \frac{1}{n-1} \left\{ n\sigma^2 - n(\overline{x}-\mu)^2 \right\} = \frac{1}{n-1} \left\{ n\sigma^2 - n\frac{1}{n}\sigma^2 \right\} \quad \text{(根據(5)式)}$$

$$= \frac{1}{n-1}(n-1)\sigma^2$$

$$= \sigma^2$$

3. 區間估計

   (1) 因點估值與被估參數間通常會有差距，故改估計一個區間，並賦予一機率以表示被估參數落在此區間之可能性。

   (2) $P(L < \mu < U)= 1 - \alpha$

      a. 區間(L, U)為被估值 μ 的「信任區間(Confidence interval)」。

      b. L 及 U 分別代表「信任下界(Lower confidence limit)」及「信任上界(Upper confidence limit)」。

      c. α 稱「顯著水準(Level of significance)」，代表 μ 不在區間(L, U)之機率。

      d. $(1 - \alpha)$稱「信任水準(Confidence level)」，代表 μ 在區間(L, U)之機率。

   (3) 信任區間愈寬，則信任水準愈高。

   (4) 點估值亦可視為信任區間寬度為 0 之區間估計值。

   (5) 例：$(90, 100)$為 μ 之 85%的信任區間，表示對 μ 落在$(90, 100)$內有 85% 的信心。

■ 3.4　抽樣－計數值抽樣
(Attribute sampling 屬性抽樣)

1.　計數值抽樣指試驗之出象僅二分為「合格」、「不合格」二種。

2.　抽樣分配(Sampling distribution)

(1)　隨機抽取 n 個樣本，其中有 k 個經測試為失效，則失效機率 P 之估計量為 $\hat{P} = \dfrac{k}{n}$，$\hat{P}$ 為點估值。

(2)　重複上述抽樣多次，可得多個 $\hat{P}$，因 $\hat{P}$ 為隨機變數，所以 $\hat{P}$ 所形成之分配應為二項式分配。$f(k) = \dbinom{n}{k} P^k \left(1 - P\right)^{n-k} = \dbinom{n}{n\hat{P}} P^{n\hat{P}} \left(1 - P\right)^{n - n\hat{P}}$，

其中 $\hat{P} = 0, \dfrac{1}{n}, \dfrac{2}{n}, ..., 1$。此離散機率函數即稱為抽樣分配函數。

(3)　$\hat{P} = \dfrac{k}{n}$ 為不偏估計量，i.e. $E\left(\hat{P}\right) = \sum \hat{P} f(\hat{P}) = \mu_{\bar{P}} = P$。

(4)　$\sigma_{\hat{P}}^{\,2} = \dfrac{1}{n} P(1 - P)$

證明：

$$\sigma_{\hat{P}}^{\,2} = V\left(\hat{P}\right) = V\left(\frac{k}{n}\right) = E\left(\frac{k}{n} - \mu_{\frac{k}{n}}\right)^2 = E\left(\frac{k}{n}\right)^2 - \left(\mu_{\frac{k}{n}}\right)^2$$

$$= \frac{1}{n^2} E\left(k^2\right) - \left(\mu_{\frac{k}{n}}\right)^2 = \frac{1}{n^2}\left[V(k) + \mu_k^{\,2}\right] - \left(\mu_{\frac{k}{n}}\right)^2$$

$$= \frac{1}{n^2}\left[npq + \left(np\right)^2\right] - \left(\frac{np}{n}\right)^2$$

$$= \frac{1}{n} pq = \frac{1}{n} p\left(1 - p\right)$$

(5)　n 愈大則估計值($\hat{P}$)與被估值(P)之差距愈小，如圖 3-8 所示。

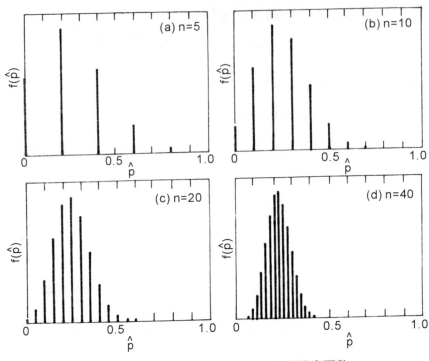

圖 3-8　不同樣本數之下的二項抽樣機率函數

3. 信任界限

(1) 設某產品之失效機率為 P，在 n 次測試中失效次數為 k。圖 3-9 所示為二項分配在不同觀測比值(k/n)以及不同失效機率(P)下的 80%、90%、95%、以及 99%信任區間。例如 95%時，$n = 10$，$k = 1 \Rightarrow$ $0.001 \leq P \leq 0.47$，可見 n 太小時所得信任區間甚大而失去估計的意義。

(2) n 愈大(則 k 必大於 0)愈有意義但耗費愈多。例如 95%時，由圖 3-9 可查得 $n=250$，$k=50 \Rightarrow$ $0.15 \leq P \leq 0.26$。

(3) 某產品之失效機率為 P，設在 n 次測試中失效次數 $k = 0$ 的機率為 γ，則 $\gamma = \binom{n}{0} P^0 (1-P)^n = (1-P)^n$，$P$ 愈大 γ 愈小。(γ 為失效會發生零次的信心水準)

(4) 若要求測試 n 次且 $k = 0$ 之機率至少為 γ，則 P 之最大值為

$$P_+ = 1 - \gamma^{\frac{1}{n}}$$ .... (3.2)

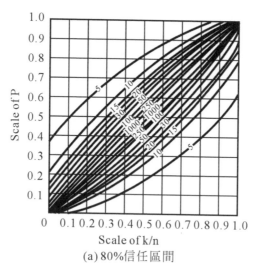

(a) 80%信任區間

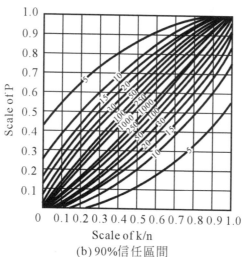

(b) 90%信任區間

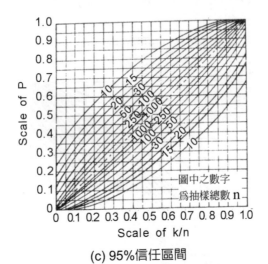

(c) 95%信任區間

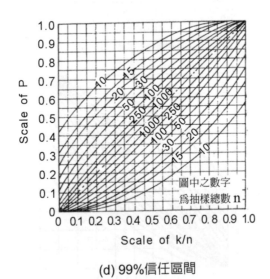

(d) 99%信任區間

圖 3-9　二項分配在不同觀測比值(*k/n*)以及不同失效機率(*P*)下的信任區間

例 3.1 一產品應測試爲無失效多少次,方能有 90%信任度證實 $p<0.1$?

解
$$P_+ = 1 - \gamma^{1/n} \implies \gamma = (1 - P_+)^n \implies \ln\gamma = n\ln(1 - P_+)$$
$$\implies n = \frac{\ln\gamma}{\ln(1-p_+)} = \frac{\ln(0.1)}{\ln(0.9)} = 21.8 \implies n = 22$$

(「90%信任度」意指相信有 90%的機率失效不會發生,會發生失效的機率只有 10%)

根據(3.2)式可將測試 n 次且測得失效次數爲 0 的特例其失效率與信任度的關係繪成圖形,如圖 3-10 所示。

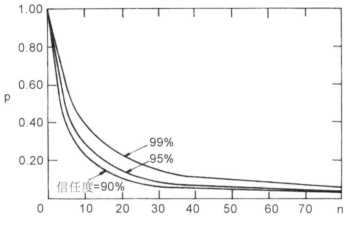

圖 3-10　測試無失效發生時失效機率的信任度

■ 3.5　驗收測試(Acceptance test)

1. 目的:對受驗品以隨機抽取方式進行檢驗,爲買方提供足夠的保證,以確認送驗批中之不良率不會高於某值。

2. 不能一一檢驗的原因:
   (1)費錢費時,
   (2)測試通常有破壞性。

3. 相關符號說明如下：

N：總受測數

n：抽樣樣本量

k：不合格數

c：允收數(Acceptance number)

P：不合格率(真值)

P_1：買方所提可接受之最大不良率

P_0：賣方認為所生產產品之最大不良率

α：生產者風險(Producer's risk)，即 $P < P_0$ 仍然被拒收的機率，通常 $\alpha \approx 5\%$

β：買方風險(Buyer's risk)，即 $P > P_1$ 仍然被接收的機率，通常 $\beta \approx 10\%$

4. 單次抽樣：一次抽樣 n 個，僅以這一次抽樣的結果判定允收或拒收。

(1) 二項分配

a. $\alpha = P(k > c \,|\, n, P_0) = \sum_{k=c+1}^{n} \binom{n}{k} P_0^{\,k} \left(1 - P_0\right)^{n-k} = 1 - \sum_{k=0}^{c} \binom{n}{k} P_0^{\,k} \left(1 - P_0\right)^{n-k}$

(指 $P < P_0$ 但因抽樣時 k 被抽到大於 c, i.e. $k = c+1, c+2, \ldots, n$, 因而仍被拒收的機率)

b. $\beta = P(k \le c \,|\, n, P_1) = \sum_{k=0}^{c} \binom{n}{k} P_1^{\,k} \left(1 - P_1\right)^{n-k}$

(指 $P > P_1$ 但因抽樣時 k 被抽到小於 c, i.e. $k = 0, 1, \ldots, c$, 因而仍被允收的機率)

(2) 波瓦松分配

使用在 n 很大而 P_0、P_1 很小時，為計算簡便，以波瓦松分配來近似二項式分配。

a. $\alpha = 1 - \sum_{k=0}^{c} \dfrac{m_0^{\,k}}{k!} e^{-m_0}$

b. $\beta = \sum_{k=0}^{c} \dfrac{m_1^{\,k}}{k!} e^{-m_1}$

c. c 及 n 的求法：

　　(a)求 $\dfrac{P_1}{P_0}$，

　　(b)由 $\dfrac{P_1}{P_0}$ 之值從表 3-4 中求出 c 值，

　　(c) $n = \dfrac{m_0}{P_0}$ 或 $n = \dfrac{m_1}{P_1}$ 。

表 3-4　$\alpha = 5\%$，$\beta = 10\%$ 時的 m_0 及 m_1 值

c	m_0	m_1	$\dfrac{P_1}{P_0}$
0	0.051	2.303	44.890
1	0.355	3.890	10.946
2	0.818	5.322	6.509
3	1.366	6.681	4.890
4	1.970	7.994	4.057
5	2.613	9.275	3.549
6	3.286	10.532	3.206
7	3.981	11.771	2.957
8	4.695	12.995	2.768
9	5.426	14.206	2.618
10	6.169	15.407	2.497
11	6.924	16.598	2.397
12	7.690	17.782	2.312
13	8.464	18.958	2.240
14	9.246	20.128	2.177
15	10.035	21.292	2.122

例 3.2 已知 $\alpha = 0.05$，$\beta = 0.1$，$P_0 = 0.02$，$P_1 = 0.05$，試決定抽樣計劃的 n、c 值。

解　$\dfrac{P_1}{P_0} = \dfrac{0.05}{0.02} = 2.5$ 由表 3-4 知　$\Rightarrow$　$c = 10$，

$n = \dfrac{m_0}{p_0} = \dfrac{6.169}{0.02} = 308.45 \approx 308$，或 $n = \dfrac{m_1}{p_1} = \dfrac{15.407}{0.05} = 308.14 \approx 308$

5. 雙次抽樣

　　相關符號說明如下：

　　N：總受測數

　　n：第一次抽樣樣本量

　　k：第一次抽樣不合格數

　　c：第一次抽樣允收數

　　r：第一次抽樣拒收數

　　m：第二次抽樣樣本量

　　q：第二次抽樣不合格數

　　d：第一次與第二次抽樣之總允收數

　　若第一次抽樣不合格數(k)雖大於允收數(c)但小於拒收數(r)，則再抽樣第二次，其流程如圖 3-11 所示。

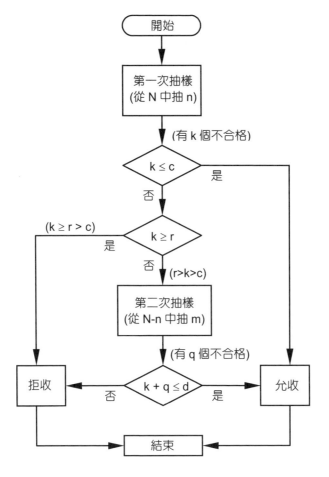

圖 3-11 雙次抽樣流程圖

■ 3.6 計量值的信任區間

1. Gamma 函數 $\Gamma(n)$

(1) 定義：$\Gamma(n) = \int_0^\infty e^{-t} t^{n-1} dt$

(2) 性質

a. $\Gamma(n+1) = n\Gamma(n)$

b. 若 $n \in N$ 則 $\Gamma(n+1) = n!$

例 3.3 已知$\Gamma(1.5) = 0.886227$，求$\Gamma(3.5)$以及$\Gamma(0.5)$。

解
$$\Gamma(3.5) = \Gamma(2.5 + 1) = 2.5\Gamma(2.5) = 2.5 \times \Gamma(1.5 + 1)$$
$$= 2.5 \times 1.5 \times \Gamma(1.5) = 2.5 \times 1.5 \times 0.886227 = 3.323351$$
$$\Gamma(1.5) = \Gamma(0.5 + 1) = 0.5\Gamma(0.5) = 0.886227$$
$$\Rightarrow \quad \Gamma(0.5) = 0.886227/0.5 = 1.772454$$

例 3.4 證明：若 $n \in N$ 則 $\Gamma(n+1) = n!$

解
$$\Gamma(1) = \int_0^\infty e^{-t} t^{1-1} dt = -e^{-t} \Big|_\infty^0 = 1$$
$$\Gamma(2) = 1 \times \Gamma(1) = 1!$$
$$\Gamma(3) = 2 \times \Gamma(2) = 2 \times 1! = 2!$$
$$\Gamma(4) = 3 \times \Gamma(3) = 3 \times 2! = 3!$$
$$\dots$$
$$\Gamma(n) = (n-1) \times \Gamma(n-1) = (n-1) \times (n-2)! = (n-1)!$$
$$\Gamma(n+1) = n!$$

2. Gamma 分配

(1) 是一種有兩個變數的分配。

(2) 常用來模擬壽命測試(Life test)的問題。

(3) $f(x) = \begin{cases} \dfrac{\lambda}{\Gamma(a)} (\lambda x)^{a-1} \exp(-\lambda x), & x \geq 0 \\ 0 & , \quad x < 0 \end{cases}$

其中λ：完全故障率

a：造成完全故障的部份故障事件數。

(4)　$\mu = \dfrac{a}{\lambda}, \quad \sigma^2 = \dfrac{a}{\lambda^2}$

當$(a-1)$為正整數(i.e. $a-1 = 1, 2, \cdots\cdots$,或 $a \geq 2$ 且 $a \in N$)時，

因 $\Gamma(a) = (a-1)!$，所以 $f(x) = \dfrac{\lambda}{(a-1)!}(\lambda x)^{a-1} \exp(-\lambda x), \quad x \geq 0$ ，

稱為 Erlangian distribution。

(5)　當 $a = 1$ 時

　　a.　$f(x) = \lambda \exp(-\lambda x), \quad x \geq 0$ ，

　　　　稱為指數分配(Exponential distribution)。

　　　　λ：Constant failure rate

　　b.　$F(x) = 1 - \exp(-\lambda x)$

　　c.　令 $\lambda = \dfrac{1}{\theta}$ ，θ 為 MTTF(不可修復)或 MTBF(可修復)則

　　　　$f(t) = \dfrac{1}{\theta}\exp(-\dfrac{t}{\theta})$ ，$F(t) = 1 - \exp(-\dfrac{t}{\theta})$

例 3.5　某型機械的失效時間呈指數分配，其平均值為 400 小時，求 8 小時內失效的機率。

解　$\theta = 400\text{h}\cdots\cdots\text{MTTF}$，$\lambda = \dfrac{1}{\theta} = \dfrac{1}{400}$ ，

$F(8) = 1 - \exp(-\dfrac{8}{400}) = 0.0198 = 1.98\%$

例 3.6　某產品之失效時間為指數分配且 MTBF = 500h，求其於 24 小時內之可靠度。

解　$R(24) = 1 - F(24) = 1 - \left[1 - \exp(-\dfrac{24}{500})\right] = \exp(-\dfrac{24}{500}) = 0.953 = 95.3\%$

---

d. 若失效時間為指數分配，則發生 x 次失效的機率為波瓦松分配。

例 3.7 某產品之失效時間為指數分配且 MTBF = 100h，求其於 1000 小時內發生 15 次失效的機率。

解 $\mu = \dfrac{1000}{100} = 10$ (1000 小時內之平均失效次數)，

$$f(x) = \frac{m^x e^{-m}}{x!}\Big|_{x=15} = \frac{10^{15} e^{-10}}{15!} = 0.0347$$

(6) 當 $\lambda = 0.5$，$a = 2k = v$ 時(v:自由度，正整數)，稱為卡方分配(Chi-square distribution)

　　a. 自由度 v、卡方值 χ^2、顯著水準 α 間之關係可由查表(附錄 H)得到。
　　　(表中之 DOF=$v=2k$，卡方值=χ_α^2)

　　b. 以常態曲線求卡方之近似值

$$\chi_\alpha^2(2k) \approx \frac{1}{2}\left(\sqrt{4k-1} + Z_{1-\alpha}\right)^2 \Rightarrow Z_{1-\alpha} \approx \sqrt{2\chi_\alpha^2(2k)} - \sqrt{4k-1}$$

　　　(其中之 $Z_{1-\alpha}$ 指常態分配表中之 $x_{F(x)}$。Z 指隨機變數，下標指 Z 以下的累積機率 $F(Z)$)

例 3.8 自由度 $v = 20$、卡方值 $\chi^2 = 33$，求顯著水準 α？

解 0.05 ：31.4104

α ：33

0.025：34.1696

以內插法可得 $\alpha=0.0368$(此為近似值，準確值須以程式計算)

例 3.9 以常態曲線近似值求 $k = 10$，$\alpha = 0.05$ 之卡方值？

解 $Z_{1-\alpha} = Z_{1-0.05} = Z_{0.95}$，即求 $P = 0.95$ 之 Z 值，查表得 $Z_{0.95} = 1.645$

∴ $\chi^2_{0.05}(20) = \dfrac{1}{2}\left(\sqrt{4 \times 10 - 1} + 1.645\right)^2 = 31.13$(查表爲 31.4104)

例 3.10 以常態曲線近似值求 $k = 10$，$\chi^2 = 33$ 之信任水準？

解 $Z_{1-\alpha} \approx \sqrt{2\chi^2_\alpha(2k)} - \sqrt{4k-1} = \sqrt{2 \times 33} - \sqrt{4 \times 10 - 1} = 1.879$，

查表可得 $1 - \alpha = 0.9698$(信任水準)

3. 產品平均壽命通常有二值須決定

(1) 信任下界(最低值)：由工程師依需求及工藝能力等因素而決定。

(2) 該信任下界之機率：由管理者依成本及商譽等因素而決定。

4. 平均壽命信任界限之測試

(1) 定時測試

a. 原則：預定一測試時數，以若干待測物進行測試，此測試執行至該預定時數爲止。

b. 雙尾信任界限：

下界：$L_{\frac{\alpha}{2},(2k+2)} = \dfrac{2 \times T}{\chi^2_{\frac{\alpha}{2}},(2k+2)}$，上界：$U_{1-\frac{\alpha}{2},2k} = \dfrac{2 \times T}{\chi^2_{1-\frac{\alpha}{2}},(2k)}$，

其中

T：所有待測物之總測試時數，

k：失效個數，

α：顯著水準，

c. 平均壽命點估值：$\bar{m} = \dfrac{T}{k}$。

例 3.11 工程師將 26 個抽風機測試 200 小時後停止，其中有 3 個失效，失效時間分別為：50 小時、61 小時、146 小時。求：

(1) 該型抽風機平均壽命的點估值？

(2) 該型抽風機平均壽命真值的 90% 信任區間？

解 $T = 200 \times 23 + 50 + 61 + 146 = 4857$（小時）

(1) 平均壽命點估值 $\bar{m} = \dfrac{T}{k} = \dfrac{4857}{3} = 1619$（小時）

(2) $\alpha/2 = 0.05$，$2k + 2 = 2 \times 3 + 2 = 8$；

$1 - (\alpha/2) = 0.95$，$2k = 2 \times 3 = 6$

$$L_{0.05,8} = \frac{2 \times 4857}{\chi^2_{0.05}(8)} = \frac{2 \times 4857}{15.5073} = 626 \text{（小時）}$$

$$U_{0.95,6} = \frac{2 \times 4857}{\chi^2_{0.95}(6)} = \frac{2 \times 4857}{1.63539} = 5940 \text{（小時）}$$

$\bar{m}$ 的 90% 信任區間為 (626, 5940) 小時

例 3.12 某產品的壽命測試總測試時間為 1650 小時，於此時間內有 10 個該產品失效。若要求信任水準為 95%，求：

(1) 平均壽命的信任雙尾？

(2) 以常態曲線近似值計算平均壽命的信任雙尾？

(3) 平均壽命點估值？

解 (1) $\chi^2_{\frac{\alpha}{2}}(2k+2) = \chi^2_{0.025}(22) = 36.781$，

$\chi^2_{1-\frac{\alpha}{2}}(2k) = \chi^2_{0.975}(20) = 9.591$

$$L_{0.025,22} = \frac{2 \times 1650}{\chi^2_{0.025}(22)} = \frac{2 \times 1650}{36.781} = 89.72 \text{（小時），}$$

$$U_{0.975,20} = \frac{2 \times 1650}{\chi^2_{0.975},(20)} = \frac{2 \times 1650}{9.591} = 344 \,(\text{小時})$$

平均壽命的信任雙尾為(89.72, 344)小時

(2) $\quad Z_{1-\frac{\alpha}{2}} = Z_{0.975} = 1.96$, $\quad Z_{\frac{\alpha}{2}} = Z_{0.025} = -1.96$

$$\chi^2_{\frac{\alpha}{2}}(2k+2) = \chi^2_{0.025}(22) = \frac{1}{2}\left(\sqrt{4 \times 11-1} + 1.96\right)^2 = 36.27 \ ,$$

$$L = \frac{2 \times 1650}{36.27} = 90.98 \,(\text{小時})$$

$$\chi^2_{1-\frac{\alpha}{2}}(2k) = \chi^2_{0.975}(20) = \frac{1}{2}\left(\sqrt{4 \times 10-1} - 1.96\right)^2 = 9.18 \ ,$$

$$U = \frac{2 \times 1650}{9.18} = 359.48 \,(\text{小時})$$

以常態曲線近似值計算平均壽命的信任雙尾為(90.98, 359.48)小時

(3) $\quad$ 平均壽命點估值 $\bar{m} = \frac{T}{k} = \frac{1650}{10} = 165 \,(\text{小時})$

---

d. 單尾(One-sided or One-tailed)信任界限

對使用者而言，通常只關心產品壽命之信任下界，即最短的壽命界限，故僅估單尾信任區間即可。

$$\bar{m} \geq L_{\alpha,(2k+2)} = \frac{2 \times T}{\chi^2_{\alpha},(2k+2)} \ , \text{如圖 3-12 所示。}$$

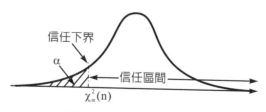

圖 3-12 單尾信任界限

例 3.13 求例 3.11 信任水準為 95% 平均壽命的信任單尾下界？

解 $\alpha = 1 - 0.95 = 0.05$，$2k + 2 = 2 \times 3 + 2 = 8$，

查表 $\chi^2_{0.05}(8) = 15.507 \Rightarrow \bar{m} \geq \dfrac{2 \times 4857}{15.5073} = 626$(小時)

與例 3.11 之結果比較，可見僅估單尾可得較高 Confidence level。

例 3.14 某產品連續操作 5000 小時均未失效，求信任水準為 95% 之平均壽命的信任單尾下界？

解 $\alpha = 1 - 0.95 = 0.05$，$k = 0$，$2k + 2 = 2$，

查表 $\chi^2_{0.05}(2) = 5.991$

$L_{0.05,2} = \dfrac{2 \times T}{\chi^2_{0.05},(2)} = \dfrac{2 \times 5000}{5.991} = 1669$ (小時)

$\Rightarrow$ 有 95% 的信心宣告該產品平均壽命的最小值為 1669 小時。

---

(2) 定數測試

    a. 原則：預定一失效個數，對若干待測物進行測試，此測試執行至發生該預定失效個數時為止。

    b. 雙尾信任界限：$L_{\frac{\alpha}{2},(2k+2)} = \dfrac{2 \times T}{\chi^2_{\frac{\alpha}{2}},(2k+2)}$，$U_{1-\frac{\alpha}{2},2k} = \dfrac{2 \times T}{\chi^2_{1-\frac{\alpha}{2}},(2k)}$，

    c. 平均壽命點估值：$\bar{m} = \dfrac{T}{k}$，

    d. λ(失效率) = 1 / MTBF(可修復) 或 λ = 1 / MTTF(不可修復)。

例 3.15 將例題 3.11 改為定數型，即測試至第 3 個失效發生即停止測試，求：

    1. 該型抽風機平均壽命的點估值？

    2. 該型抽風機平均壽命真值的 90% 信任區間？

解　　$T = 146 \times 24 + 50 + 61 = 3615$(小時)

(1)　平均壽命點估值 $\bar{m} = \dfrac{T}{k} = \dfrac{3615}{3} = 1205$ (小時)

(2)　$\alpha/2 = 0.05$，$2k + 2 = 2 \times 3 + 2 = 8$；$1 - (\alpha/2) = 0.95$，$2k = 2 \times 3 = 6$

$$L_{0.05,8} = \frac{2 \times 3615}{\chi^2_{0.05},(8)} = \frac{2 \times 3615}{15.5073} = 466.23 \text{ (小時)}$$

$$U_{0.95,6} = \frac{2 \times 3615}{\chi^2_{0.95},(6)} = \frac{2 \times 3615}{1.63539} = 4420.96 \text{ (小時)}$$

$\bar{m}$ 的 90%信任區間為(466.23, 4420.96)小時

例 3.16　同例題 3.11 但改為將 26 個抽風機測試 146 小時後停止，求：

1.　該型抽風機平均壽命的點估值？

2.　該型抽風機平均壽命真值的 90%信任區間？

解　　$T = 146 \times 24 + 50 + 61 = 3615$(小時)

(1)　平均壽命點估值 $\bar{m} = \dfrac{T}{k} = \dfrac{3615}{3} = 1205$ (小時)

(2)　$\alpha/2 = 0.05$，$2k + 2 = 2 \times 3 + 2 = 8$；$1 - (\alpha/2) = 0.95$，$2k = 2 \times 3 = 6$

$$L_{0.05,8} = \frac{2 \times 3615}{\chi^2_{0.05},(8)} = \frac{2 \times 3615}{15.5073} = 466.23 \text{ (小時)}$$

$$U_{0.95,6} = \frac{2 \times 3615}{\chi^2_{0.95},(6)} = \frac{2 \times 3615}{1.63539} = 4420.96 \text{ (小時)}$$

$\bar{m}$ 的 90%信任區間為(466.23, 4420.96)小時

　　由例題 3.11、3.15 及 3.16 可知，若定時型設定的測試時數較定數型設定第 k 個失效的時數為大時，可得較大估測結果(例題 3.11 結果較例題 3.15 結果為大)。為何？

例 3.17 某產品進行壽命測試，當第 10 個該產品失效即終止測試。若當第 10 個該產品失效時之總測試時間為 7500 小時，求：

1. 失效率真值信任水準為 90%的信任區間？
2. 平均壽命的點估值？

解

1. $1 - 1(\alpha/2) = 1 - 10.05 = 0.95$，$2k = 2 \times 10 = 20$，

$\chi^2_{0.95}(20) = 10.851$，$\chi^2_{0.05}(22) = 33.9244$

$U_{0.95,20} = \dfrac{2 \times T}{\chi^2_{0.95},(20)} = \dfrac{2 \times 7500}{10.851} = 1382$（小時）

$L_{0.05,22} = \dfrac{2 \times T}{\chi^2_{0.05},(22)} = \dfrac{2 \times 7500}{33.9244} = 442$（小時）

$\Big\}$ 平均壽命的信任區間為(442, 1382)小時

而失效率 $\lambda = \dfrac{1}{MTBF}$

$\bar{m} = 442 \Rightarrow \lambda = \dfrac{1}{442} = 0.002262$（次/小時）

$\bar{m} = 1382 \Rightarrow \lambda = \dfrac{1}{1382} = 0.000723$（次/小時）

$\Big\}$ 失效率真值的信任區間為 (0.000723, 0.002262) 次/小時

2. 平均壽命點估值 $\bar{m} = \dfrac{T}{k} = \dfrac{7500}{10} = 750$（小時）

5. 測試時數之決定，以下列數例說明之。

例 3.18 某產品要求其壽命至少為 100 小時，對其進行壽命測試時，第一次失效發生即停止測試。若要求該產品之生產有 95%符合要求，問該測試應進行之最少時數？其平均壽命點估值為多少小時？又若取100 個該產品測試至第 10 個失效發生時的總測試時數為 4500 小時，問該批產品是否合於要求？

解　以定時型之單尾信任界限來討論。$\bar{m} \geq L_{\alpha,(2k+2)} = \dfrac{2 \times T}{\chi^2_\alpha,(2k+2)}$

1.　$L_{0.05,(2\times1+2)} = \dfrac{2 \times T}{\chi^2_{0.05},(4)} = \dfrac{2T}{9.48773} = 100 \Rightarrow T = 474.39$ 小時

2.　$\bar{m} = \dfrac{T}{k} = \dfrac{474.39}{1} = 474.39$ (小時)

3.　$L_{0.05,(2\times10+2)} = \dfrac{2 \times 4500}{\chi^2_{0.05},(22)} = \dfrac{2 \times 4500}{33.92443} = 265.30 > 100$

（預估最低壽命合於要求）

$\bar{m} = \dfrac{T}{k} = \dfrac{4500}{10} = 450 > 100$ （平均壽命點估值亦合於要求）

例 3.19　某產品要求其MTBF之點估值至少為100小時，現對50個該產品進行壽命測試，若有失效則立即修復或更換至第10次失效發生時為止。求(1)測試總時數，(2)最低壽命95%信心之預估值，(3)若要求最低壽命95%信心之預估值至少為100小時，測試總時數至少須為多少？

解　1.　平均壽命點估值 $\bar{m} = \dfrac{T}{k} = \dfrac{T}{10} = 100 \Rightarrow T = 1000$ 小時

2.　$L_{0.05,(2\times10+2)} = \dfrac{2 \times 1000}{\chi^2_{0.05},(22)} = \dfrac{2 \times 1000}{33.92443} = 58.95$ (小時)

3.　$L_{0.05,(2\times10+2)} = \dfrac{2 \times T}{\chi^2_{0.05},(22)} = \dfrac{2 \times T}{33.92443} = 100$ (小時)，$T = 1696.22$ 小時

6.　由要求條件求信任水準，以下列數例說明之。

例 3.20　現對某產品進行壽命測試，若至第 10 次失效發生時之總測試時數為 1650 小時。現若要求該產品須至少操作 100 小時而無失效，求達成此要求之信任水準？

解 $L_{\alpha,(2\times10+2)} = \dfrac{2\times1650}{\chi_\alpha^2,(22)} = 100 \Rightarrow \chi_\alpha^2,(22) = 33$

$0.05:33.9244 \qquad \alpha:33 \qquad 0.1:30.8133$

以內插法可得 $\alpha = 0.0649$ (此為近似值，準確值須以程式計算)

信任水準為 $1-\alpha = 1-0.0649 = 0.9351 = 93.51\%$

---

7. 求倖存機率

倖存機率(Survival probability) P：在期望壽命下的可靠度，或是可達成期望壽命的機率。

$P = e^{-\frac{t}{m}}$，

$\overline{m}$：平均壽命點估值(即可靠度隨時間呈指數遞減之時間常數，失效個數為 0 時，以單尾信任下界取代)，

t：期望壽命。

例 3.21 某產品 100 個均連續操作 1000 小時且均未失效，根據此測試結果及信任水準為 90% 的信任單尾下界求該產品完成 1000 小時任務的倖存機率？

解 $T = 100\times1000 = 10^5$ (小時)，$\alpha = 1-0.9 = 0.1$，

$k = 0$，$2k+2 = 2$，查表 $\chi_{0.1}^2(2) = 4.60517$

$\overline{m} \geq L_{\alpha,2k+2} = \dfrac{2T}{\chi_{0.1,(2\times0+2)}^2} = \dfrac{2\times10^5}{4.60517} = 43431$ (小時)

倖存機率 $p = e^{-\frac{t}{m}} = e^{-\frac{1000}{43431}} = e^{-0.023} = 0.9773$

$\Rightarrow$ 有 90% 的信心宣告該產品完成 1000 小時任務的倖存機率至少為 97.73%

8.　失效率非定值而呈常態分配時的情形

(1)　失效率之抽樣呈常態分配

a.　若母體呈常態分配 $N(\mu, \sigma^2)$，則可證明由其中抽取之隨機樣本的平均數 $\bar{x}$ 亦呈常態分配 $N(\bar{x}, \dfrac{\sigma^2}{n})$，其中 $\dfrac{\sigma^2}{n} = \sigma_{\bar{x}}^2$。

b.　$\sigma_{\bar{x}}$：樣本平均數的標準差，亦稱爲估計值的標準誤(Standard error of the estimate)。

c.　信任區間

(a) 設顯著水準爲 α，因標準常態分配左右對稱於 0，故其信任區域 $(1-\alpha)$ 的上、下限(i.e.估計壽命的上下限)勢必爲 $Z_{\frac{\alpha}{2}}$ 與 $Z_{1-\frac{\alpha}{2}}$，

故 $P\left(Z_{\frac{\alpha}{2}} < \dfrac{\bar{x} - \mu}{\sigma_{\bar{x}}} < Z_{1-\frac{\alpha}{2}} \right) = 1 - \alpha$，

其中 $Z_{\frac{\alpha}{2}}$ 與 $Z_{1-\frac{\alpha}{2}}$ 分別表示標準常態隨機變數 Z 於 $\dfrac{\alpha}{2}$ 百分位數 (percentile，指累積分配機率)及 $(1 - \dfrac{\alpha}{2})$ 百分位數所對應之隨機變數，如圖 3-13 所示。

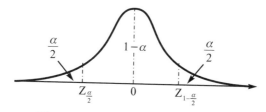

圖 3-13　標準常態分配的信任區間

(b) 單尾：$Z_\alpha = \dfrac{L_\alpha - \bar{x}}{\sigma_{\bar{x}}} \Rightarrow L_\alpha = \bar{x} + Z_\alpha \sigma_{\bar{x}}$

d.　作標準常態分配轉換時，$Z = \dfrac{\bar{x} - \mu}{\sigma_{\bar{x}}}$。

e.　若樣本量 $n \geq 30$，則雖母體分布未知，但依 CLT 可知 $\bar{x}$ 的抽樣分配爲常態分配。

例 3.22 由抽樣數據估計某機器的壽命可以常態分配 $N(100, 10^2)$ 描述，單位為月。在管理者要求信任水準為 90%的條件下，求該機器壽命的信任下界。

解 $\alpha = 1 - 0.9 = 0.1$，$L_\alpha = \bar{x} + Z_\alpha \sigma_{\bar{x}}$

$L_{0.1} = 100 + Z_{0.1} \times (10) = 100 + (-1.282) \times (10) = 87.18\,(月)$

例 3.23 由抽樣數據估計某機器的壽命可以常態分配 $N(100, 10^2)$ 描述，單位為月。若要求該機器壽命的信任下界為 80 個月，求信任水準。

解 $L_\alpha = \bar{x} + Z_\alpha \sigma_{\bar{x}} \Rightarrow Z_\alpha = \dfrac{L_\alpha - \bar{x}}{\sigma_{\bar{x}}} = \dfrac{80 - 100}{10} = -2.0$

$\Rightarrow$ 查表得 $\alpha = 0.0228$

$1 - \alpha = 1 - 0.0228 = 0.9772 = 97.72\%$

例 3.24 若管理者要求在信任水準為 90%下某機器壽命之信任下界為 100，標準誤是 10，單位為月。求平均壽命。(以 L_a 代替 $\bar{x}$)

解 $\alpha = 0.1$，$L_{0.1} = 100$，$\sigma_{\bar{x}} = 10$，$Z_{0.1} = -1.282$

$Z_\alpha = \dfrac{\bar{x} - \mu}{\sigma_{\bar{x}}} = \dfrac{L_\alpha - \mu}{\sigma_{\bar{x}}} \Rightarrow Z_{0.1} = \dfrac{L_{0.1} - \mu}{\sigma_{\bar{x}}} \Rightarrow -1.282 = \dfrac{100 - \mu}{10}$

$\Rightarrow \mu = 112.82$ 月

(2) 失效率之抽樣呈 t 分配

    a. 信任區間：$L_\alpha = \bar{x} + \left[t_{\frac{\alpha}{2}}, (n-1) \right] \sigma_{\bar{x}}$，$U_\alpha = \bar{x} + \left[t_{1-\frac{\alpha}{2}}, (n-1) \right] \sigma_{\bar{x}}$

    b. 單尾下界：$t_{\alpha,(n-1)} = \dfrac{L_{\alpha,n} - \bar{x}}{\sigma_{\bar{x}}}$ (標準化)$\Rightarrow L_{\alpha,n} = \bar{x} + \left[t_\alpha, (n-1) \right] \sigma_{\bar{x}}$

c. t 分配爲常態分配於小樣本時之修正分配，$n<30$ 時均採 t 分配

d. $n-1=\nu$(自由度)

e. 以 $t_{\alpha,}(\nu)$ 表示隨機變數 T 大於 $t_{\alpha,}(\nu)$ 時(自由度ν)之機率爲α，即

　　$P\!\left(T>t_{\alpha,}(\nu)\right)=\alpha$。

f. t 分配左右對稱，i.e. $P\!\left(T<-t_{\alpha,}(\nu)\right)=P\!\left(T>t_{\alpha,}(\nu)\right)$，或

　　$-t_{\alpha,}(\nu)=t_{1-\alpha,}(\nu)$，如圖 3-14 所示。

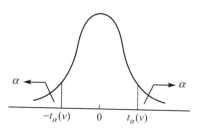

圖 3-14　t 分配的信任區間

例 3.25　由抽樣數據估計某機器的壽命可以常態分配 $N(100,\ 10^2)$描述，該數據係由樣本量 $n=10$ 求得，單位爲小時。(1)現管理者要求信任下界爲 80 小時，求信任水準；(2)若管理者要求信任下界爲 120 小時，求顯著水準。

解

(1) $L_{\alpha,n}=\bar{x}+\left[t_{\alpha,}(n-1)\right]\sigma_{\bar{x}}\ \Rightarrow\ 80=100+\left[t_{\alpha,}(10-1)\right]10$

　　$\Rightarrow t_{\alpha,}(9)=\dfrac{80-100}{10}=-2$

　　現欲求 $\left[t_{\alpha,}(9)=-2\right]=1-\left[t_{\alpha,}(9)=2\right]$，查 t 分布數表(附錄 I)以及使用內插法，得 $\left[t_{\alpha,}(9)=2\right]$之$\alpha=0.0403$，

　　故信任水準爲 $1-\alpha=1-0.0403=0.9597=95.97\%$

(Note：相較例題 3.23，本題爲小樣本，故信心水準較低(95.97% < 97.72%))

(2) $L_{\alpha,n}=\bar{x}+\left[t_{\alpha,}(n-1)\right]\sigma_{\bar{x}}\ \Rightarrow\ 120=100+\left[t_{\alpha,}(10-1)\right]10$

　　$\Rightarrow t_{\alpha,}(9)=\dfrac{120-100}{10}=2$

現欲求 $\left[t_\alpha (9) = 2 \right]$，查 t 分布數表以及使用內插法，

得 $\left[t_\alpha (9) = 2 \right]$ 之 $\alpha = 0.0403$，

故顯著水準為 $\alpha = 0.0403 = 4.03\%$

---

9. 可靠度的信任界限

a. $R_{LC} = e^{-\frac{t}{L_\alpha}}$：指定信任下界之下的可靠度(Reliability under the lower confidence limit)

b. $R_{UC} = e^{-\frac{t}{U_\alpha}}$：指定信任上界之下的可靠度(Reliability under the upper confidence limit)

其中

L_α：平均壽命的單尾下界，U_α：平均壽命的上界，

t：期望的任務時間。

例 3.26 已知 $L_\alpha = 105$ 小時之信任水準為 95%，求

1. 該機器 10 小時可靠度的信任下界，
2. 100 小時之 R_{LC}。

解
1. $R_{LC} = e^{-\frac{10}{105}} = 0.9092$，換句話說，$P\left(R_{10} \geq 0.9092 \right) = 95\%$
2. $R_{LC} = e^{-\frac{100}{105}} = 0.3858$，換句話說，$P\left(R_{100} \geq 0.3858 \right) = 95\%$

例 3.27 已知 $L_\alpha = 95.87$ 小時，$U_\alpha = 353$ 小時，其信任水準為 95%，求該機器 10 小時可靠度的信任界限。

解 $R_{LC} = e^{-\frac{10}{95.87}} = 0.9009$，$R_{UC} = e^{-\frac{10}{353}} = 0.9721$

$\Rightarrow$ 有 95%的信心宣告該產品 10 小時可靠度的信任界限為(90.09%, 97.21%)

❖ 習 題

1. 某工程師執行馬達壽命測試，執行測試之樣品數為 26，其中前三個失效時間分別為第 50，61，146 小時。

   (1) 若上述數據係執行定時測試 200 小時所得之結果，求：

      (a) 平均壽命眞值的 95%信任區間，

      (b) 平均壽命眞值的點估值，

      (c) 若僅估信任單尾，其值為多少小時？

   (2) 若上述數據係執行定數測試所得之結果，且該測試執行至第三個樣品失效即停止，求：

      (a) 其平均失效率 95%的信任區間為何？

      (b) 平均壽命眞值的點估值？

   (3) 前述二種測試，其中一種所估得之壽命長於另一種，原因為何？

2. 某工程師執行馬達壽命測試，執行測試之樣品數為 100，由開始至第 1500 小時始有第一個馬達失效。求：

   (1) 95%信任的 MTBF 值？

   (2) 該馬達於 2000 小時內的可靠度？

3. 某產品的失效時間是依指數分布，其平均值為 400 小時。(1)求 10 小時內之可靠度？(2)若 MTBF=500h，該產品在 24 小時內之失效率是多少？(3)若 MTBF=100h，求在 1000 小時內發生 10 次失效的機率？

4. 某機器連續操作 10000 小時均無發生失效，求平均壽命之 95%信任水準的單尾下界。

5. 要求某產品之平均壽命至少為 100 小時，現對其進行壽命測試，且要求第一次出現失效即停止測試。若期望該產品有 95%符合要求，則

   (1) 該測試應進行多少小時？

   (2) 其平均壽命如何？

   (3) 若取 100 個該產品測試至第 10 個失效出現所花之總測試時間為 5000 小時，該批產品是否合格？

Chapter *4*

可靠度與失效分析

■ 4.1 失效與失效率曲線

1. 失效(Failure)：產品未能發揮預定機能的狀態。包括：

   (1) 故障

   (2) 不穩定

   (3) 功能減退

   誤差(Error)、失效(Failure)、故障(Fault)三者間的關係示於圖 4-1。此三者之定義如下[IEC, 1990]：

   (1) Error is a discrepancy between a computed, observed or measured value or condition and the true, specified or theoretically correct value or condition.

   (2) Failure is an event when a required function is terminated (exceeding the acceptable limits).

(3) Fault is the state characterized by inability to perform a required function, excluding the inability during preventive maintenance or other planned actions or due to lack of external resources.

　　根據以上的定義可知，誤差還不是失效，而故障是由失效所導致的狀態。有時候誤差被稱為「初步的失效(Incipient failure)」，所以預防保養(Preventive maintenance)的執行應在系統的行為(Performance)還在可接受變異範圍(Acceptable deviation)內之誤差狀態下實施。

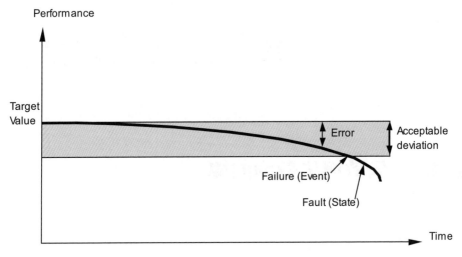

圖 4-1　誤差(Error)、失效(Failure)、故障(Fault)三者間的關係

2. 失效的分類

　　(1) 依原因

　　　　a. 誤用失效(Misuse failure)

　　　　b. 先天性弱點失效(Inherent weakness failure)

　　(2) 依情況

　　　　a. 突發性失效(Sudden failure)

　　　　b. 漸次性失效(Gradual failure)

(3) 依程度

    a.　局部失效(Partial failure)

    b.　全面失效(Complete failure)

(4) 依組合

    a.　致命性失效(Catastrophic failure)：突發性失效 ＋ 全面失效

    b.　退化性失效(Degradation failure)：漸次性失效 ＋ 局部失效

(5) 依性質

    a.　主要失效(Primary failure)：隨機或自然發生

    b.　次級失效(Secondary failure)：過度壓力(Over stress)造成

    c.　指令失效(Command failure)：錯誤控制所造成

    d.　共同模式失效(Common mode failure)：共同原因所造成

3. 失效的來源

(1) 老化(Aging)

(2) 人員(Human)

(3) 環境(Environment)

(4) 組件(Component)

4. 壽命曲線

(1) 可靠度也稱為：「時間維度的品質(Quality in time dimension)」，也就是說可靠度會隨時間而改變。

(2) 浴缸曲線(Bathtub curve)

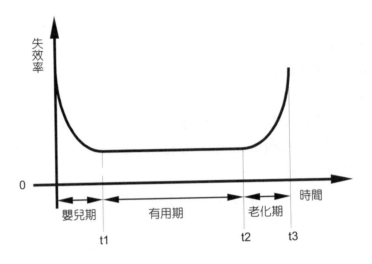

圖 4-2　浴缸曲線

　　大部分產品甚至包括人類的壽命曲線都近似浴缸曲線，也就是在壽命的早期有隨時間遞減的失效率(如人類的新生兒沒有自我照護的能力，且抵抗力低，故經常生病。但隨著年紀漸長，自我照護的能力和抵抗力增加，故生病的頻率逐漸下降)、壽命的中期有最低且幾乎是定值的失效率(如人類在青壯年時不常生病)、壽命的末期則失效率隨時間遞增(如人類的晚年因老化等因素使得生病的頻率又逐漸上升)。其中轉折的關鍵時間即為圖 4-2 中的 t1 及 t2，兩者間的間距為壽命的有用期。然不同開發程度國家的人民會因衛生條件、醫療水準、社福制度等等因素而呈現差異但類似的壽命曲線。開發中國家人民的壽命曲線較高且較圓滑，代表了較高的生病頻率(失效率)、較不明顯的有用期、較短的平均壽命，如圖 4-3 所示。浴缸曲線可以分為三個部分，描述如下。

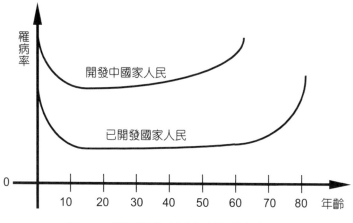

圖 4-3　不同開發度國家人民的壽命曲線

a. 早夭期(Early failure period)；或稱「除錯期(Debugging phase)」；
   或稱「嬰兒期(Infant mortality)」：
   失效主要來自先天缺陷，可能原因有：
   (a) 設計上的失誤
   (b) 製程中的失誤
   (c) 檢驗時的失誤
   (d) 使用時的失誤
   (e) 其他如儲存、運輸、包裝上的失誤
       主要藉「環境應力篩選(Environment Stress Screening, ESS)」
       來剔除。

b. 機遇(隨機)失效期（Chance failure period）；或稱「使用期（Useful
   life）」：
   失效主要來自意外事故或疾病，可能原因有：
   (a) 安全係數太低
   (b) 不當外力
   (c) 早夭期未發作的潛伏性缺失
   (d) 人為失誤
   (e) 濫用

(f) 天災或環境

(g) 其他無法解釋的原因

主要藉「足夠設計或足夠安全邊限(Safety margin)」來降低其發生機率。

c. 磨耗期(Wear-out period)；或稱「老化期(Aging period)」：

失效主要來長期累積的原因，計有：

(a) 磨耗

(b) 疲乏

(c) 腐蝕

(d) 脆裂

(e) 其他如短壽命設計或不當翻修

主要藉「全員生產管理(Total Productive Management, TPM)」來延緩其發生的時間。

(3) 不同產品的壽命曲線

a. 電子硬體：如圖 4-4，重隨機失效

b. 電腦軟體：如圖 4-5，重除錯

c. 機械設備：如圖 4-2，重預防保養

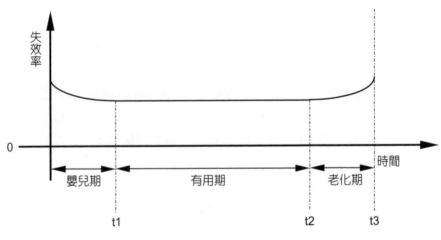

圖 4-4　電子硬體的壽命曲線

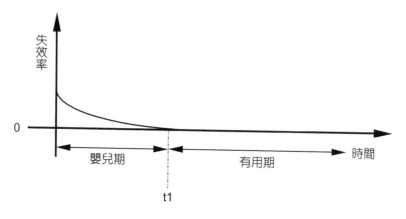

圖 4-5　電腦軟體的壽命曲線

(4)　壽命曲線類型依 t1 及 t2 不同而不同。

■ 4.2　可靠度函數

1.　令 $P(T)$ 為失效發生在時間 T 的機率(Probability)，若 T 介於 t 與 $t+\Delta t$ 之間則 $P(T)=P(t \leq T \leq t+\Delta t)$。

2.　失效機率密度函數(Failure probability density function)：

$$f(t) = \frac{P(t \leq T \leq t + \Delta t)}{\Delta t}$$

3.　失效機率累積函數(Failure probability cumulative function)：$F(t) = P(T \leq t)$
指所有發生在小於或等於時間 t 之失效機率的和，i.e. $F(t) = \int_0^t f(t)dt$，

如圖 4-6 所示。

4.　可靠度函數：$R(t) = P(T > t)$
指所有失效發生在時間 t 之後(或說在 t 之前不會失效)之機率的和，i.e.
$R(t) = \int_t^\infty f(t)dt$，如圖 4-6 所示。

$R(0) = 1$，$R(\infty) = 0$，$R(t) = 1 - F(t) = 1 - \int_0^t f(t)dt$　$\Rightarrow$　$f(t) = -\dfrac{d}{dt}R(t)$

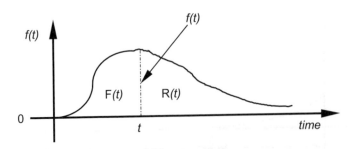

圖 4-6　失效機率累積函數與可靠度函數

5.　危險函數(Hazard function)或瞬間失效率函數(Instantaneous failure rate)：$\lambda(t)$

$\lambda(t) = \dfrac{P(T < t + \Delta t \,|\, T > t)}{\Delta t}$，指在 $T > t$ 之前提下，將失效發生在 $T < t + \Delta t$

的機率平均分配在時間Δt 內所得的失效率(單位時間內的失效機率)。

$\because P(X|Y) = \dfrac{P(X \cap Y)}{P(Y)}$

$\therefore \lambda(t)\Delta t = P(T < t + \Delta t \,|\, T > t) = \dfrac{P\big[(T < t + \Delta t) \cap (T > t)\big]}{P(T > t)}$

$= \dfrac{P(t < T < t + \Delta t)}{P(T > t)} = \dfrac{f(t)\Delta t}{R(t)} \quad \Rightarrow \quad \lambda(t) = \dfrac{f(t)}{R(t)}$

例 4.1　假設某產品壽命的機率密度函數為 $f(t) = 0.25te^{-0.5t}$，單位為年。求：

1.　該產品第一年內失效的機率？

2.　該產品至少可維持五年不失效的機率？

3.　期望不超過 5%的該產品需要保固服務(Warranty service)，則該產品的保固期至多可設為多少個月？

解　$F(t) = \displaystyle\int_0^t 0.25te^{-0.5t}\,dt = 1 - (1 + 0.5t)\,e^{-0.5t}$

1.　$F(1) = 1 - (1 + 0.5 \times 1)\,e^{-0.5 \times 1} = 0.0902$

2.　$P(T > t) = 1 - P(T \le t) = 1 - F(t) = R(t)$

　　$R(5) = 1 - F(5) = 1 - \big[1 - (1 + 0.5 \times 5)\,e^{-0.5 \times 5}\big] = 0.2873$

3. 設可保固 t_0 年，由題意可知 $P(T > t_0) \geq 0.95$，

　而現 $R(1) = 1 - F(1) = 1 - 0.0902 = 0.9098$，故 t_0 必定小於一年。

　試 $t_0 = 6/12$，$P(T > 0.5) = R(0.5) = 0.973$

　$\Rightarrow$　$F(0.5) = 1 - 0.973 = 0.027 = 2.7\%$

　試 $t_0 = 9/12$，$P(T > 0.75) = R(0.75) = 0.945$

　$\Rightarrow$　$F(0.75) = 1 - 0.945 = 0.055 = 5.5\%$

　試 $t_0 = 8/12$，$P(T > \dfrac{8}{12}) = R(\dfrac{2}{3}) = 0.955$

　$\Rightarrow$　$F(\dfrac{2}{3}) = 1 - 0.955 = 0.045 = 4.5\%$

　故 t_0 最多可設為 8 個月。

---

6. 平均值定理(Mean Value Theorem, MVT)

　若 $f(t)$ 為連續函數，則

　$P(a \leq T \leq b) = \displaystyle\int_a^b f(t)dt = (b-a)f(\xi)$，其中 ξ 為介於 a 與 b 中間之數，且

　$\xi = a + \mathrm{m}(b-a)$, $0 < \mathrm{m} < 1$。

　若 $(b-a) = \Delta t \approx 0$ (i.e. very small)，則 $t^* = \dfrac{a+b}{2} \approx \xi$

　$\Rightarrow$　$P(a \leq T \leq b) = \displaystyle\int_a^b f(t)dt = (b-a)f(t^*) = \Delta t\, f(t^*)$。

例 4.2 N(100, 625)，請以 MVT 法求 $P(104 \leq T \leq 106)$？

解　$\mu = 100$，$\sigma = 25$

$P(104 \leq T \leq 106) = (106 - 104)f(\dfrac{104 + 106}{2}) = 2 \times f(105)$

$$= 2 \times \frac{1}{\sqrt{2\pi}\,(25)} \exp\left[-\frac{1}{2}\left(\frac{105 - 100}{25}\right)^2\right] = 0.03128$$

另解：

$$= P(104 \le T \le 106) = P\left(\frac{104-100}{25} \le Z \le \frac{106-100}{25}\right) = P(0.16 \le Z \le 0.24)$$

$$= P(Z \le 0.24) - P(Z \le 0.16) = 0.5948 - 0.5636 = 0.03120$$

例 4.3 某組件的壽命機率密度函數為 $f(t) = 0.002e^{-0.002t}$（小時），求失效介於 510 與 515 小時間的機率？

1. 以 MVT 法求近似值，
2. 以積分法求精確值？

解

1. $f\left(\dfrac{510+515}{2}\right)(515-510) = 0.002e^{-0.002(512.5)} \times 5 = 0.00358796$

2. $F(515) - F(510) = \displaystyle\int_0^{515} 0.002e^{-0.002t}\,dt - \int_0^{510} 0.002e^{-0.002t}\,dt$

   $= e^{-0.002(510)} - e^{-0.002(515)} = 0.00358798$

7. $\lambda(t)$、$R(t)$、$f(t)$ 間之關係

   (1) $\lambda(t) = \dfrac{f(t)}{R(t)} = \dfrac{1}{R(t)} \times \left[-\dfrac{d}{dt}R(t)\right]$

   $\left(\because R(t) = 1 - F(t) = 1 - \displaystyle\int_0^t f(t)\,dt \quad \Rightarrow \quad f(t) = -\dfrac{d}{dt}R(t)\right)$

   $\Rightarrow \quad \lambda(t)dt = \dfrac{1}{R(t)} \times [-dR(t)] \quad \Rightarrow \quad \int \lambda(t)dt = -\ln[R(t)]$

   $\Rightarrow \quad R(t) = \exp\left[-\displaystyle\int_0^t \lambda(t)dt\right]$（以 $\lambda(t)$ 表示 $R(t)$）

   (2) $f(t) = \lambda(t) \times R(t) = \lambda(t) \times \exp\left[-\displaystyle\int_0^t \lambda(t)dt\right]$（以 $\lambda(t)$ 表示 $f(t)$）

8. $\text{MTTF} = \displaystyle\int_0^\infty tf(t)dt = -\int_0^\infty t \times \dfrac{dR(t)}{dt}dt = -\left[\int_0^\infty t\,dR(t)\right]$

   $= -\left[tR(t)\Big|_0^\infty - \displaystyle\int_0^\infty R(t)dt\right] = \int_0^\infty R(t)dt$

例 **4.4** 假設 $R(t) = \begin{cases} \left(1 - \dfrac{t}{t_0}\right)^2 & 0 \le t < t_0 \\ 0 & t \ge t_0 \end{cases}$ ，求：

1. $\lambda(t)$，

2. 判斷 $\lambda(t)$ 隨 t 遞增或遞減，

3. MTTF。

解

1. $f(t) = -\dfrac{d}{dt}R(t) = -\dfrac{d}{dt}\left(1 - \dfrac{t}{t_0}\right)^2 = \dfrac{2}{t_0}\left(1 - \dfrac{t}{t_0}\right)$ ， $0 \le t < t_0$

$$\lambda(t) = \frac{f(t)}{R(t)} = \frac{\dfrac{2}{t_0}\left(1 - \dfrac{t}{t_0}\right)}{\left(1 - \dfrac{t}{t_0}\right)^2} = \frac{2}{t_0\left(1 - \dfrac{t}{t_0}\right)} \quad , \quad 0 \le t < t_0$$

$\lambda(t) = 0$ ， $t \ge t_0$

2. $\left.\begin{array}{l} \lambda(0) = \dfrac{2}{t_0} \\ \lambda(t_0) = \infty \end{array}\right\}$ ∴ $\lambda(t)$ 隨 t 遞增。

3. $\text{MTTF} = \displaystyle\int_0^{t_0}\left(1 - \dfrac{t}{t_0}\right)^2 dt = \int_0^{t_0}\left(1 - 2\dfrac{t}{t_0} + \dfrac{t^2}{t_0^2}\right)dt$

$= \left. t \right|_0^{t_0} - \dfrac{2}{t_0} \times \dfrac{1}{2}t^2 \Big|_0^{t_0} + \dfrac{1}{t_0^2} \times \dfrac{1}{3}t^3 \Big|_0^{t_0}$

$= t_0 - \dfrac{1}{t_0}\left(t_0^2\right) + \dfrac{1}{3t_0^2}\left(t_0^3\right) = \dfrac{1}{3}t_0$ ， $0 \le t < t_0$

■ 4.3 隨機失效(Random failure)

1. 特性

(1) 失效機率與使用年限及過去之操作歷程無關。

(2) 對「需求失效(Demand failure)」而言，失效機率與過去已需求之次數無關。

(3) 對連續操作系統而言，其具有定值失效率(Constant failure rate)。

(4) 常採指數分配來描述。

2. 指數分配

(1) $f(t) = \lambda e^{-\lambda t}$

$F(t) = 1 - e^{-\lambda t}$

$R(t) = e^{-\lambda t}$

其中 t 指已 debug 之後，wear-out 之前，在有用壽命(Useful time)之內的時間，$\lambda(t) = \lambda$ 為定值常數。

MTBF(或 MTTF)$= \dfrac{1}{\lambda}$，$\sigma^2 = \dfrac{1}{\lambda^2}$。

例 4.5　某裝置有定值失效率 $\lambda = 0.02$ 次/小時，求：

1. 該裝置在操作開始之後 10 小時之內失效的機率？

2. 假設該裝置已不失效地操作 100 小時，求自第 101 小時起，10 小時之內失效的機率？

解

1. $F(10) = P(T \le 10) = \int_0^{10} f(t)dt = 1 - e^{-0.02 \times 10} = 0.181$

2. $P(T \le 110 \,|\, T > 100)$

$= \dfrac{P[(T \le 110) \cap (T > 100)]}{P(T > 100)} = \dfrac{P(100 \le T \le 110)}{P(T > 100)}$

$= \dfrac{\int_{100}^{110} f(t)dt}{1 - F(100)} = \dfrac{\int_{100}^{110} 0.02 e^{-0.02t} dt}{1 - (1 - e^{-0.02 \times 100})} = 0.181$

因失效率為定值，故任意 10 小時內的失效機率均相等。

例 4.6 一已 debug 之系統的 useful time 為 1000h，$\lambda = 0.0001/h$。求：

1. 在此 1000h 內任 10h 之 $R(t)$？
2. 在此 1000h 內之 $R(t)$？

解

1. $R(10) = e^{-0.0001 \times 10} = 0.9990$ (即使在 991~1000 之 10h 內的不失效機率仍為 99.9%)

2. $R(1000) = e^{-0.0001 \times 1000} = 0.9048$ (有 90.48% 之機率可操作 1000h 而不失效)

---

(2)　$R(t) = e^{-\frac{t}{m}}$，t 為任務時間(Mission time)而非曆時間(Calendar time)，

$t = 0$ 指任務開始的時間。$t = m$，$R(t) = e^{-\frac{m}{m}} = e^{-1} = 0.368$

$\Rightarrow \quad m = \text{MTBF} = \tau$ (Time-constant of reliability)

例 4.7 假設要求某系統於 1h 之內的可靠度為 0.9999。求：

1. MTBF？
2. 操作 10h 內之 R(t)？

解

1. $R(t) = e^{-\frac{t}{m}} \quad \Rightarrow \quad 0.9999 = e^{-\frac{1}{m}} \quad \Rightarrow \quad m = 10000\,\text{h}$
2. $R(10) = e^{-\frac{10}{10000}} = 0.999$

---

(3)　組件數 n 與系統 MTBF 間的關係

　　a.　若系統由 n 個相同組件組成，每一組件之 MTBF $= m$。若其中任一組件失效則系統失效，該系統之 MTBF $= \dfrac{m}{n}$。

　　b.　若各組件之 MTBF 不同，則該系統之 MTBF $= \dfrac{\overline{m}}{n}$，$\overline{m}$ 為組件 MTBF 之平均數。

c. 以 m 爲橫座標之可靠度曲線稱標準化可靠度曲線(Standardized curve)

d. 可靠度曲線之橫座標參數亦可由時間改爲可描述該系統特性之參數，如開關之開閉次數(Cycle)。

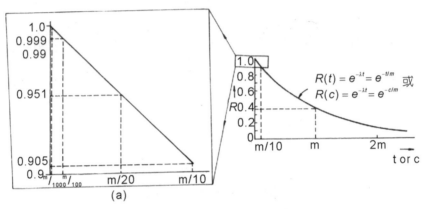

圖 4-7　標準化可靠度曲線

例 4.8 設某物品之 MTBF $= m$。求：

1. 該品操作 $t = \dfrac{m}{1000}$ 時間之可靠度？

2. 一系統含 1000 個該品，該系統操作 $\dfrac{m}{1000}$ 時間內之 $R(t)$？

3. 一系統含 10 個該品，該系統操作 $\dfrac{m}{1000}$ 時間內之 $R(t)$？

解

1. $R(t) = e^{-\frac{t}{m}} = e^{-\frac{\frac{m}{1000}}{m}} = e^{-\frac{1}{1000}} = 0.999$

   解讀：a. 1 個該品操作 $\dfrac{m}{1000}$ 時間內，失效機率爲 0.001 或 0.1%。

   b. 1000 個該品操作 $\dfrac{m}{1000}$ 時間內，僅有一個可能失效。

2. 1000 個該品操作 $\dfrac{m}{1000}$ 時間內，均不失效的機率(可靠度)為

$$R(t) = e^{-\frac{t}{\left(\frac{m}{n}\right)}} = e^{-\frac{\frac{m}{1000}}{\left(\frac{m}{1000}\right)}} = e^{-1} = 0.368 \ (1000 \text{ 個該品的 MTBF} = \frac{m}{1000})$$

3. $R(t) = e^{-\frac{t}{\left(\frac{m}{n}\right)}} = e^{-\frac{\frac{m}{1000}}{\left(\frac{m}{10}\right)}} = e^{-\frac{1}{100}} = 0.99 \ (10 \text{ 個該品的 MTBF} = \frac{m}{10})$

(4) 可靠度 R 與系統 MTBF 間的關係

$R(t) = e^{-\frac{t}{m}}$，現設 $t = 1$，$m = 5$ $\Rightarrow$ $R = 82\%$；

$m = 10$ $\Rightarrow$ $R = 90\%$，MTBF 增長一倍，可靠度增加 8%；

$m = 20$ $\Rightarrow$ $R = 95\%$，MTBF 再增長一倍，可靠度僅增加 5%；

$m = 100$ $\Rightarrow$ $R = 99\%$，MTBF 再增長五倍，但可靠度僅增加 4%。(邊際效用遞減)

(5) MTBF 可作為安排維修率(Repair rate)及備品(Spare parts)個數的參考。

(6) 某些設備的 MTBF 概略值如表 4-1 所示。

表 4-1　某些設備的 MTBF 概略值

設　備	MTBF 概略值
海底電纜的 Repeat amplifier	20 年
人造衛星(部份)	10~20 年
電腦及實驗室電子設備	5000~10000 小時
軍用地面器材	1000~5000 小時
船艦電子器材	500~2500 小時
空用電子器材	100~1000 小時
飛彈電子器材	1~500 小時
電梯	5 年或 44000 小時
電視機	5000~10000 小時
電冰箱	10000 小時
洗衣機	5000~10000 小時

3. 複合失效率(Composite failure rate)

複合失效率(λ)：為運轉(Operating)失效率(λ_0)、需求(Demand)失效率(P)、備用狀態(Standby)失效率(λ_S)的總和。

$$\lambda = \frac{c}{t}P + c\lambda_0 + (1-c)\lambda_S \quad\text{.. (4.1)}$$

其中 $\overline{t} = \dfrac{T(累積運轉時間)}{m(需求次數)}$：平均運轉時間，

$c = \dfrac{T}{t}$：時間能力因子(Capacity factor)或稱「任務週期(Duty cycle)」，

t：可靠度或失效率要求的曆時間。

(4.1)式中λ為 P、λ_O、λ_S 三者的線性組合，該三者的係數代表了各自在複合失效率(λ)中的權重。其中 P 的係數 $\dfrac{c}{t} = \dfrac{\dfrac{T}{t}}{\dfrac{T}{m}} = \dfrac{m}{t}$ 表示在可靠度或失效率要求的曆時間中的需求頻率。

例 4.9 某泵浦為間歇式運作，該泵浦的運轉失效率為 0.0004 次/小時，備用狀態失效率為 0.00001 次/小時，需求失效率為 0.0005 次/小時。在過去 24 小時該馬達的啟動時間(up-time, t_u)及停機時間(down-time, t_d)如下表所示，求：

1. 在過去 24 小時該馬達的複合失效率？
2. 若這些數據有代表性，30 天內該馬達的失效機率？

t_u	*0.78*	1.69	*2.89*	3.92	*4.71*	5.97	*6.84*	7.76
t_d	*1.02*	2.11	*3.07*	4.21	*5.08*	6.31	*7.23*	8.12
t_u	8.91	*9.81*	10.81	*11.87*	12.98	*13.81*	14.87	*15.97*
t_d	9.14	*10.08*	11.02	*12.14*	13.18	*14.06*	15.19	*16.09*
t_u	*16.69*	17.71	*18.61*	19.61	*20.56*	21.49	*22.58*	23.61
t_d	*16.98*	18.04	*19.01*	19.97	*20.91*	21.86	*22.79*	23.89

解　由題目知：$\lambda_0 = 0.0004$，$P = 0.0005$，$\lambda_S = 0.00001$，$m = 24$，所以

$$T_d = \sum_{i=1}^{24} t_{di} = 301.5(hr)，T_u = \sum_{i=1}^{24} t_{ui} = 294.36(hr)$$

$$T = T_d - T_u = 301.5 - 294.36 = 7.14(hr)$$

$$\bar{t} = \frac{7.14}{24} = 0.2975(hr)，c = \frac{7.14}{24}。$$

1.　$\lambda = \dfrac{0.2975}{0.2975} \times 0.0005 + 0.2975 \times 0.0004 + (1 - 0.2975) \times 0.00001$

　　$= 0.000626$ (次/hr)

2.　$R(t) = e^{-\lambda t} = e^{-(0.000626 \times 24 \times 30)} = 0.637$

　　$F(t) = 1 - R(t) = 1 - 0.637 = 0.363 = 36.3\%$

■ 4.4　早夭失效率與使用期失效率之關係

設 T_0 為燒入測試執行的總時間(Burn-in period)，對通過燒入測試之裝置而言，即為已順利運轉的時間；

t 為使用時間(Useful time)，即通過燒入測試後可再倖存時間；

$T = (T_0 + t)$ 為由新品開始至失效所需的時間；

其間關係如圖 4-8 所示。

圖 4-8　時間 T_0、t、$T_0 + t$ 間的關係

令 $R(t|T_0)$ 為通過燒入測試後之使用期的可靠度（Reliability of useful time），則

$$R(t|T_0) = P(T > T_0 + t \,|\, T > T_0) = \frac{P(T > T_0 + t) \cap P(T > T_0)}{P(T > T_0)}$$

$$= \frac{P(T > T_0 + t)}{P(T > T_0)} = \frac{R(T_0 + t)}{R(T_0)} = \frac{\exp\left[-\int_0^{T_0+t} \lambda(x)dx\right]}{\exp\left[-\int_0^{T_0} \lambda(x)dx\right]} = \exp\left[-\int_{T_0}^{T_0+t} \lambda(x)dx\right]$$

$$\Rightarrow \quad \frac{d}{dT_0} R(t|T_0) = \exp\left[-\int_{T_0}^{T_0+t} \lambda(x)dx\right] \times \frac{d}{dT_0}\left[-\int_{T_0}^{T_0+t} \lambda(x)dx\right]$$

$$= R(t|T_0) \times \left\{-\left[\lambda(T_0 + t) - \lambda(T_0)\right]\right\}$$

$$= R(t|T_0) \times \left[\lambda(T_0) - \lambda(T_0 + t)\right]$$

所以，若 $\lambda(T_0) > \lambda(T_0 + t)$（早夭失效率大於使用期失效率，i.e.失效率隨時間遞減），則 $R(t|T_0)$ 之斜率為正，也就是說 $R(t|T_0)$ 隨燒入測試執行的總時間增加而變大。故燒入測試有助於提升使用期的可靠度。

■ 4.5 以特定分配描述失效率

1. 常態分配(Normal distribution)

   (1) 失效機率密度函數 $f(t) = \frac{1}{\sigma\sqrt{2\pi}} \exp\left[-\frac{1}{2}\left(\frac{t-\mu}{\sigma}\right)^2\right]$，$\mu = \text{MTTF}$

   失效累積分配函數 $F(t) = \int_{-\infty}^{t} f(t)dt$

   標準化後之失效累積分配函數 $\Phi\left(\frac{t-\mu}{\sigma}\right)$

   可靠度 $R(t) = 1 - \Phi\left(\frac{t-\mu}{\sigma}\right)$

   失效率 $\lambda(t) = \frac{f(t)}{R(t)}$

(2)　信任區間百分比

表 4-2　標準常態分配之信任區間百分比

標準差 σ	信任區間
±0.5	0.3830
±1.0	0.6826
±1.5	0.8664
±1.645	0.9000
±2.0	0.9544
±2.5	0.9876
±3.0	0.9973

(3)　適用於損耗時間 μ 相當明確的場合。

例 4.10　某型輪胎已知有 90% 的機率在行駛於 25000 哩至 35000 哩之間損耗報廢，若該型輪胎壽命為常態分配，求：1. μ？　2. σ？

解

解法一：$\left.\begin{array}{l}\mu - \Delta t = 25000 \\ \mu + \Delta t = 35000\end{array}\right\} \Rightarrow \begin{array}{l}\mu = 30000 \\ \Delta t = 5000\end{array}$

$\because P(\mu - \Delta t \leq T \leq \mu + \Delta t) = 90\%$ ，

$\therefore \Delta t = 1.645\sigma \Rightarrow 5000 = 1.645\sigma \Rightarrow \sigma = 3039$ 哩

解法二：$\because P(25000 \leq T \leq 35000) = 0.9$ ，

$\therefore \left.\begin{array}{l}\Phi\left(\dfrac{25000 - \mu}{\sigma}\right) = 0.05 \\[2mm] \Phi\left(\dfrac{35000 - \mu}{\sigma}\right) = 0.95\end{array}\right\}$ 查表得 $\begin{array}{l}\Phi(-1.645) = 0.05 \\ \Phi(1.645) = 0.95\end{array}$

$\left.\begin{array}{l}\dfrac{25000 - \mu}{\sigma} = -1.645 \\[2mm] \dfrac{35000 - \mu}{\sigma} = 1.645\end{array}\right\}$ 解得 $\begin{array}{l}\mu = 30000 \\ \sigma = 3039\end{array}$

2. 對數常態分配(Lognormal distribution)

(1) 常用於描述疲乏、老化或損耗導致失效率隨時間而增大的現象，尤其適用於 TTF 有相當大不確定性的情形。

(2) $f(t) = \dfrac{1}{st\sqrt{2\pi}} \exp\left\{ -\dfrac{1}{2s^2}\left[\ln\left(\dfrac{t}{t_0}\right) \right]^2 \right\}$

$\text{MTTF} = \mu = t_0 \exp\left(\dfrac{s^2}{2}\right)$

$\sigma^2 = t_0^2 \exp\left(s^2\right)\left[\exp\left(s^2\right) - 1 \right]$

(3) 若 90% 之信任區間為 $\left[t_- = \dfrac{t_0}{n}, t_+ = t_0 n \right]$ 則 $t_0 = \left(t_- \times t_+\right)^{\frac{1}{2}}$，$s = \dfrac{1}{1.645}\ln(n)$

例 4.11 某型貨車輪軸已知在行駛於 120,000 哩至 180,000 哩之間疲乏失效，若該型貨車輪軸壽命為對數常態分配，求：

1. 信任度 90% 之下的因子 n？
2. 參數 t_0 及 s？
3. MTTF？
4. 標準差？

解

1. $t_0 n = t_+ = 180,000$，$t_0/n = t_- = 120,000$

   $\Rightarrow n^2 = \dfrac{180,000}{120,000} \Rightarrow n = 1.2247$

2. $t_0^2 = 180,000 \times 120,000 \Rightarrow t_0 = 146.97 \times 10^3$

   $\Rightarrow s = \dfrac{1}{1.645}\ln(n) = \dfrac{\ln(1.2247)}{1.645} = 0.1232$

3. $\text{MTTF} = t_0 \exp\left(\dfrac{s^2}{2}\right) = 146.97 \times 10^3 \times \exp(0.1232^2 / 2)$

   $= 148.09 \times 10^3 \,(\text{哩})$

4. $\sigma^2 = t_0^2 \exp\left(s^2\right)\left[\exp\left(s^2\right) - 1\right]$

$= \left(146.97 \times 10^3\right)^2 \times \exp\left(0.1232\right)^2 \times \left[\exp\left(0.1232\right)^2 - 1\right]$

$= 335,403,102$

$\sigma = \sqrt{335,403,102} = 18314$ (哩)

---

3. 韋氏分配(Weibull distribution)

(1) 二參數型式

　　a. 失效率：$\lambda(t) = \dfrac{\beta}{\alpha}\left(\dfrac{t}{\alpha}\right)^{\beta-1}$ ，

　　　　α：尺度參數(Scale parameter)，$\alpha > 0$。
　　　　β：形狀參數(Shape parameter)，$\beta > 0$。

　　　　可靠度：$R(t) = \exp\left[-\int_0^t \lambda(t)\,dt\right] = \exp\left[-\left(\dfrac{t}{\alpha}\right)^{\beta}\right]$

　　　　失效機率密度函數：$f(t) = \lambda(t)R(t) = \dfrac{\beta}{\alpha}\left(\dfrac{t}{\alpha}\right)^{\beta-1} \times \exp\left[-\left(\dfrac{t}{\alpha}\right)^{\beta}\right]$

　　　　失效累積分配函數：$F(t) = 1 - R(t) = 1 - \exp\left[-\left(\dfrac{t}{\alpha}\right)^{\beta}\right]$

　　b. $\mu = \text{MTTF} = \alpha \times \Gamma\left(1 + \dfrac{1}{\beta}\right)$ 　$\sigma^2 = \alpha^2\left\{\Gamma\left(1 + \dfrac{2}{\beta}\right) - \left[\Gamma\left(1 + \dfrac{1}{\beta}\right)\right]^2\right\}$

　　c. $\beta = 1$ 　$f(t) = \dfrac{1}{\alpha}\exp\left(-\dfrac{t}{\alpha}\right)$ ，即為 Exponential distribution。

　　d. $\beta = 2$ 　$f(t) = \dfrac{2t}{\alpha^2}\exp\left[-\left(\dfrac{t}{\alpha}\right)^2\right]$ ，即為 Rayleigh distribution。

　　e. $\beta = 1$ 可描述隨機失效的定值失效率，
　　　　$\beta > 1$ 可描述失效率遞增的老化效應，
　　　　$\beta < 1$ 可描述失效率遞減的燒入效應。

　　f. 經過共執行 T_0 燒入測試時間之後的可靠度：$R\left(t\middle|T_0\right) = \dfrac{R\left(T_0 + t\right)}{R\left(T_0\right)}$ 。

g. 當 α(尺度參數)固定時，韋氏分配的形狀會隨 β(形狀參數)而變。當 $\beta = 1$ 時，韋氏分配即為指數分配；當 $\beta \geq 3$ 時，韋氏分配會類似常態分配。請見圖 4-9。

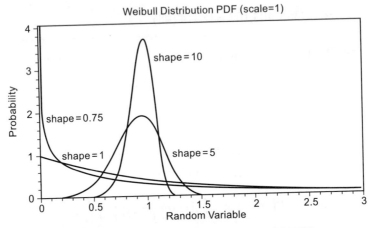

圖4-9　不同 β(形狀參數)時韋氏分配的形狀

h. 因 $\mu = \alpha \times \Gamma\left(1 + \dfrac{1}{\beta}\right)$，當 $\beta \geq 1$ 時，$\Gamma\left(1 + \dfrac{1}{\beta}\right) \approx 1$，即 $\mu \approx \alpha$。所以知道 α 即知道分配的中心位置(尤其是當 $\beta \geq 3$ 時)。請見圖 4-10。

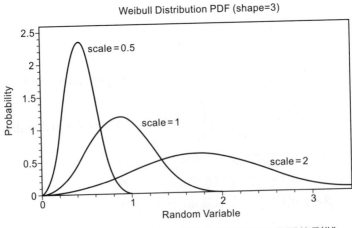

圖4-10　$\beta = 3$ 時，不同 α(尺度參數)下韋氏分配的形狀

例 **4.12** 某組件的 TTF 可用 Weibull 分配描述，其二參數 $\alpha = 1000$ 小時，
$\beta = 2$，求：

1.　該組件操作 200 小時的可靠度？
2.　該組件的 MTTF？

解

1.　$R(t) = \exp\left[-\left(\dfrac{t}{\alpha}\right)^{\beta}\right]$ $\Rightarrow$ $R(200) = \exp\left[-\left(\dfrac{200}{1000}\right)^{2}\right] = 0.9608$

2.　$\text{MTTF} = \alpha\Gamma\left(1 + \dfrac{1}{\beta}\right) = 1000\Gamma\left(1 + \dfrac{1}{2}\right) = 1000\Gamma(1.5) = 886.23\,(\text{小時})$

例 **4.13** 假設某組件有一遞減失效率，可用 Weibull 分配描述，其二參數
$\alpha = 180$ 年，$\beta = 1/2$。若已知該組件的設計壽命可靠度為 0.9，求：

1.　若沒有經過燒入測試，其設計壽命多長？
2.　若該組件啟用前經過一個月的燒入測試，其設計壽命多長？

解

1.　$R(t) = \exp\left[-\left(\dfrac{t}{\alpha}\right)^{\beta}\right]$

$\Rightarrow$ $t = \alpha\left\{\ln\left[\dfrac{1}{R(t)}\right]\right\}^{1/\beta} = 180\left[\ln\left(\dfrac{1}{0.9}\right)\right]^{2} = 2\,(\text{年})$

2.　經過燒入 T_0 時間後的可靠度為：

$R(T \mid T_0) = \dfrac{\exp\left[-\left(\dfrac{t + T_0}{\alpha}\right)^{\beta}\right]}{\exp\left[-\left(\dfrac{T_0}{\alpha}\right)^{\beta}\right]}$ $\Rightarrow$ $t = \alpha\left\{\ln\left[\dfrac{1}{R(t)}\right] + \left(\dfrac{T_0}{\alpha}\right)^{\beta}\right\}^{1/\beta} - T_0$

$= 180\left[\ln\left(\dfrac{1}{0.9}\right) + \left(\dfrac{1}{12 \times 180}\right)^{1/2}\right]^{2} - \dfrac{1}{12} = 2.81\,(\text{年})$

(2) 三參數型式

    a. 較二參數型式多了一個參數 t_0，適用於描述先經過一段時間(t_0)後才會失效的現象。

只需將二參數型式之時間軸原點向右移動 t_0 即可。稱為位置參數(Location parameter)或免失效參數(Failure-free time)，或最低壽命(Minimum life)。

    b. 失效率：$\lambda(t) = \begin{cases} 0 & ,t < t_0 \\ \dfrac{\beta}{\alpha}\left(\dfrac{t}{\alpha}\right)^{\beta-1} & ,t \geq t_0 \end{cases}$

可靠度：$R(t) = \begin{cases} 1 & ,t < t_0 \\ \exp\left[-\left(\dfrac{t}{\alpha}\right)^{\beta}\right] & ,t \geq t_0 \end{cases}$

失效機率密度函數：$f(t) = \begin{cases} 0 & ,t < t_0 \\ \dfrac{\beta}{\alpha}\left(\dfrac{t}{\alpha}\right)^{\beta-1} \times \exp\left[-\left(\dfrac{t}{\alpha}\right)^{\beta}\right] & ,t \geq t_0 \end{cases}$

失效累積分配函數：$F(t) = \begin{cases} 0 & ,t < t_0 \\ 1 - \exp\left[-\left(\dfrac{t}{\alpha}\right)^{\beta}\right] & ,t \geq t_0 \end{cases}$

$\mu = \text{MTTF} = \alpha \times \Gamma\left(1 + \dfrac{1}{\beta}\right) + t_0$ (向右移動 t_0)

$\sigma^2 = \alpha^2\left\{\Gamma\left(1 + \dfrac{2}{\beta}\right) - \left[\Gamma\left(1 + \dfrac{1}{\beta}\right)\right]^2\right\}$ (與二參數型式同)

■ 4.6　可用度(Availability)

1. 至失效時間(Time To Failure, TTF)

   由修復剛完成(或新品)起，至失效剛發生時為止的時段，亦即操作時間或上線時間(Uptime)。

2. 至修復時間(Time To Repair, TTR)

   由失效剛發生時起，至修復剛完成時為止的時段，亦即停機時間或下線時間(Downtime)。包括檢查、待料、修理……等時間的總和。

3. 失效間隔(Time Between Failure, TBF)

   由前一次失效剛發生時起，至本次失效剛發生時為止的時段，亦即一次downtime 加一次 uptime。

4. 修復間隔(Time Between Repair, TBR)

   由前一次修復剛完成時(或新品)起，至本次修復剛完成時為止的時段，亦即一次 uptime 加一次 downtime。

5. 可用度 A(t)：在指定時間內，上線時間與該指定時間的比值，亦即

$$A(t) = \frac{\sum_{i=1}^{n}(\text{TTF})_i}{\sum_{i=1}^{n}\left[(\text{TTF})_i + (\text{TTR})_i\right]}$$

例 **4.14** 請根據表 4-3 求：

　　1. A(t)？　　2. MTTF？　　3. MTTR？　　4. MTBF？　　5. MTBR？

　　6. Mean Failure Rate？　　7. Mean Repair Rate？

表 4-3　TTF/TTR 之記錄表

TTF(小時)	TTR(小時)
125	1.0
44	1.0
27	9.8
53	1.0
8	1.2
46	0.2
5	3.0
20	0.3
15	3.1
12	1.5
58	1.0
53	0.8
36	0.5
25	1.7
106	3.6
200	6.0
159	1.5
4	2.5
29	0.3
27	3.8
小計 1102	49.8
總計 1151.8	

解　設 F_i 指第 i 次失效剛發生的時間、R_i 指第 i 次修復剛完成的時間，其中 $i = 1, 2, \cdots\cdots, 20$。則 TTF、TTR、TBF、TBR 等間之關係如圖 4-11 所示。

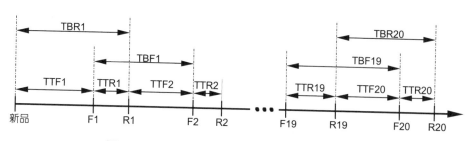

圖 4-11　TTF、TTR、TBF、TBR 等間之關係

1. $A(t) = \dfrac{1102}{1151.8} = 0.95676$

2. $MTTF = \dfrac{1102}{20} = 55.10\,(\text{小時})$ （Mean uptime）

3. $MTTR = \dfrac{49.8}{20} = 2.49\,(\text{小時})$ （Mean downtime）

4. $MTBF = \dfrac{1151.8 - 125 - 3.8}{19} = 53.84\,(\text{小時})$

5. $MTBR = \dfrac{1151.8}{20} = 57.59\,(\text{小時})$

6. $\text{Mean Failure Rate } \lambda = \dfrac{1}{MTBF} = \dfrac{1}{53.84} = 0.01857\,(\text{次/小時})$

7. $\text{Mean Repair Rate } \gamma = \dfrac{1}{MTBR} = \dfrac{1}{57.59} = 0.01736\,(\text{次/小時})$

❖ 習 題

1. 請敘述浴缸曲線(Bath-tub curve)之特性，並說明每階段失效發生的原因。

2. 請繪出描述產品失效率的浴缸曲線，並說明

   (1) 曲線中分哪三個時期？

   (2) 每個時期失效的主因？

   (3) 如何改善每個時期的失效率？

3. 某產品之可靠度函數為 $R(t) = \begin{cases} (1-\dfrac{t}{t_0})^2, 0 \leq t < t_0 \\ 0, t \geq t_0 \end{cases}$ ，求：

   (1) MTTF，

   (2) 失效率 $\lambda(t)$，

   (3) $\lambda(t)$ 為時間之遞增或遞減函數？

4. 某產品之失效函數為 $f(t) = 0.01te^{-0.01t}, t \geq 0$，其中 t 之單位為年。

   (1) 求第二年內之失效機率？

   (2) 求五年內之可靠度？

   (3) 若提供保固期(Warranty)為九個月，預估將有多少百分比的產品接受保固服務？

5. 某組件之壽命機率密度函數為 $f(t) = 0.01te^{-0.01t}, t \geq 0$，其中 t 之單位為小時。求失效介於 100 與 102 小時間的機率？

   (1) 以 MVT 法求近似值，

   (2) 以積分法求精確值。

6. 設 $\lambda(t)$ 為失效率函數，$R(t)$ 為可靠度函數，$f(t)$ 為失效機率密度函數，求：

   (1) 寫出兩兩之間的關係式，

   (2) 以 $R(t)$ 及 $f(t)$ 表示出 MTTF，

(3) 若 $R(t) = \begin{cases} (1 - \dfrac{t}{t_0})^2, 0 \le t < t_0 \\ 0, t \ge t_0 \end{cases}$ ，求 $\lambda(t)$ 以及 MTTF。

7. 某社區大門設有自動照明設備，凡有人員經過則自動照明一分鐘後停止。該自動照明設備的運轉失效率為每小時 0.00001 次，備用狀態失效率為每小時 0.000001 次，需求失效率為每小時 0.00002 次，且此系統之供電時間為每天下午六點至次日清晨六點。

   (1) 若該社區平均每小時有十人次進出該大門，求該自動照明設備在此系統之供電時間內的複合失效率為多少？

   (2) 24 小時內的複合失效率為多少？

   (3) 一個月(30 天)內的可靠度為多少？

   (4) 一年(365 天)內的可靠度為多少？

8. 某機車製造商估計該公司之機車有 90%的機率在 25000 公里至 35000 公里之間報廢。

   (1) 若該型機車的壽命為常態分配，求該型機車的：

      (a) 平均壽命，

      (b) 標準差。

   (2) 若該型機車的壽命為對數常態分配，求該型機車的 MTTF。

9. 某型機具底座之減震器已知有 90%在使用於 12,000 小時至 16,000 小時之間老化失效，若該型機具底座之減震器壽命為對數常態分配，求：

   (1) 期望值，

   (2) 標準差。

10. 某產品過去 10 次 TTF 和 TTR 記錄如下表，求：

   (1) Availability：$A(t)$，

   (2) MTTF，

(3) MTTR，

(4) MTBF，

(5) MTBR。

TTF(小時)	TTR(小時)
125	1.0
44	1.0
27	9.8
53	1.0
8	1.2
46	0.2
5	3.0
20	0.3
15	3.1
12	1.5

Chapter 5

可靠度評估

■ 5.1 緒言

可靠度評估(Reliability evaluation)在美軍規格的定義是：「利用物品內的零件、機能、操作環境以及彼此間的相互關係等知識，評估該物品未來可靠度動向的技術」。

■ 5.2 靜態評估模式(Static evaluation model)

靜態模式是指「系統中任一組件的失效不影響其他組件的功能，亦即組件間失效為彼此獨立」。

1. 串聯系統(Serial system)的可靠度

    (1) 定義：由獨立單元所連結而成的複雜系統，其中任一單元失效將會導致整個系統失效的組件連結方式。

    (2) 該系統的可靠度受其最弱單元可靠度的限制。

(3) 該系統的可靠度爲各單元可靠度的乘積，亦即：$R_S = R_1 \times R_2 \times ... \times R_N$

(4) 若 R_S 保持不變，當組件數 N 越多時，則各組件可靠度必須提高，或說各組件的失效率必須相對降低。

(5) 若不採取補償措施，則當一系統的複雜度增加(亦即組件數增多)，其系統可靠度必然下降。

(6) 若各組件之可靠度相同爲(R_C)，則：$R_S = (R_C)^N$。

若各組件之可靠度不相同，則組件之平均可靠度：$R_C = \sqrt[N]{R_1 \times R_2 \times ... \times R_N}$，

亦即爲幾何平均。

(7) 系統可靠度與串聯組件數的關係曲線

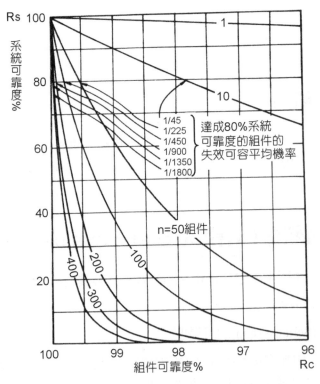

圖 5-1 系統可靠度與串聯組件數的關係曲線

例 5.1　某產品由三個組件串聯構成，其可靠度分別為 0.92，0.95，0.96，求
該產品之可靠度？

解　$R_S = R_1 \times R_2 \times R_3 = 0.92 \times 0.95 \times 0.96 = 0.839 = 83.9\%$

例 5.2　某產品由三個次系統串聯構成，次系統 1 由 20 個可靠度為 0.99 的零
件串聯而成；次系統 2 由 10 個可靠度為 0.98 的零件串聯而成；次系
統 3 由 30 個可靠度為 0.999 的零件串聯而成。求該產品之可靠度？

解　$R_S = 0.99^{20} \times 0.98^{10} \times 0.999^{30}$

$= 0.8179 \times 0.817 \times 0.97 = 0.648 = 64.8\%$

例 5.3　某產品由四個組件串聯構成，其失效率分別為 0.002，0.001，0.0025，
0.0005。求該產品操作 100 小時之可靠度？

解　$R_{100} = e^{-0.002 \times 100} \times e^{-0.001 \times 100} \times e^{-0.0025 \times 100} \times e^{-0.0005 \times 100}$

$= 0.8187 \times 0.9048 \times 0.7788 \times 0.9512 = 0.5488$

例 5.4　假設某電路由 4 個電晶體(失效率 0.00001 次/小時)、10 個二極體(失
效率 0.000002 次/小時)、20 個電阻(失效率 0.000001 次/小時)、10 個
電容(失效率 0.000002 次/小時)串聯構成，求：

1.　該電路之 MTBF，

2.　該電路十小時內之可靠度？

解　1.　$\sum \lambda = 4(0.00001) + 10(0.000002)$

$+ 20(0.000001) + 10(0.000002) = 0.0001$

$\text{MTBF} = \dfrac{1}{0.0001} = 10,000(小時)$

2.　$R(10) = e^{-0.0001 \times 10} = 0.999$

2. 並聯系統(Parallel system)的可靠度

  (1) 定義：僅當所有組件均失效時，系統方才失效的組件連結方式。又稱「主動複聯(Active redundancy)或熱複聯(Hot redundancy)」。

  (2) 另有待命並聯系統(Standby parallel system)：系統中有多於一個具有相同機能的組件，當正在操作中的組件失效時，立即可由備用組件(Backup component)取代而維持系統於不失效的設計。又稱「被動複聯(Passive redundancy)或冷複聯(Cold redundancy)」。

  (3) 兩組件並聯

    a. 設 X 表組件可操作的事件，則系統可靠度可表示如下：

$$R = P(X_1 \cup X_2) = P(X_1) + P(X_2)$$
$$-P(X_1 \cap X_2) = R_1 + R_2 - R_1 \times R_2 \quad\text{.......................(5.1)}$$

    b. 設 λ 表組件之定值失效率，則系統可靠度可表示如下：

$$R(t) = e^{-\lambda_1 t} + e^{-\lambda_2 t} - e^{-\lambda_1 t} \times e^{-\lambda_2 t} = e^{-\lambda_1 t} + e^{-\lambda_2 t} - e^{-(\lambda_1 t + \lambda_2 t)} \quad\text{....................(5.2)}$$

    c. $\text{MTTF} = \int_0^\infty R(t)dt = \dfrac{1}{\lambda_1} + \dfrac{1}{\lambda_2} - \dfrac{1}{\lambda_1 + \lambda_2}$ ....(5.3)

    d. 若 $\lambda_1 = \lambda_2 = \lambda$ 則

$$R(t) = 2e^{-\lambda t} - e^{-2\lambda t} \quad\text{..(5.4)}$$

$$\text{MTTF} = \dfrac{2}{\lambda} - \dfrac{1}{2\lambda} = \dfrac{3}{2\lambda} \quad\text{..(5.5)}$$

    e. 若單一組件失效率為定值 λ，則二組件並聯系統之失效率 λ_s 可推導如下：

$$\because \lambda(t) = \frac{f(t)}{R(t)} = \frac{1}{R(t)} \times \left[-\frac{d}{dt}R(t) \right] ,$$

    由(5.4)式 $R(t) = 2e^{-\lambda t} - e^{-2\lambda t}$，代入上式，得

$$\lambda_s(t) = \lambda \left(\frac{1 - e^{-\lambda t}}{1 - 0.5e^{-\lambda t}} \right) \quad\text{...} (5.6)$$

失效率為定值之單一組件與二組件並聯系統之可靠度及失效率隨時間之變化示於圖 5-2。由(5.6)式及圖 5-2 可知：

(a) 雖單一組件失效率為定值，但並聯系統之失效率卻為時間相依(Time dependent)。

(b) 並聯系統之失效率由零開始隨時間($t = 0$ 時 $R = 1$)逐漸遞增，與老化期之失效率曲線相似，其間差異有二：

    (i) 並聯系統之失效率曲線為凹向下(Concave down)

    (ii) 並聯系統之失效率曲線以定值失效率為漸進上限(Asymptotic upper-limit)

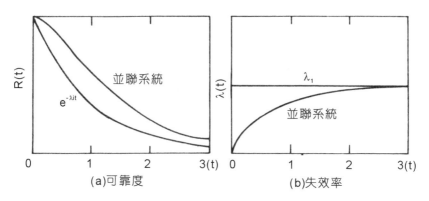

圖 5-2　單一組件與二組件並聯系統之可靠度及失效率

(4) 三組件並聯

a.　$R(t) = 1 - (1 - e^{-\lambda_1 t}) \times (1 - e^{-\lambda_2 t}) \times (1 - e^{-\lambda_3 t})$ .... (5.7)

b.　$\text{MTTF} = \dfrac{1}{\lambda_1} + \dfrac{1}{\lambda_2} + \dfrac{1}{\lambda_3} - \dfrac{1}{\lambda_1 + \lambda_2}$

$$- \dfrac{1}{\lambda_2 + \lambda_3} - \dfrac{1}{\lambda_3 + \lambda_1} + \dfrac{1}{\lambda_1 + \lambda_2 + \lambda_3} \quad\text{.................................} (5.8)$$

c. 若 $\lambda_1 = \lambda_2 = \lambda_3 = \lambda$ 則

$$R(t) = 1 - (1 - e^{-\lambda t})^3 = 3e^{-\lambda t} - 3e^{-2\lambda t} - e^{-3\lambda t} \dots\dots\dots\dots\dots\dots (5.9)$$

$$\text{MTTF} = \frac{3}{\lambda} - \frac{3}{2\lambda} + \frac{1}{3\lambda} = \frac{11}{6\lambda} \dots\dots\dots\dots\dots\dots\dots (5.10)$$

(5) n 組件並聯

    a. $R(t) = 1 - (1 - e^{-\lambda_1 t}) \times (1 - e^{-\lambda_2 t}) \times \cdots\cdots \times (1 - e^{-\lambda_n t})$

$$= 1 - \prod_{i=1}^{n} (1 - e^{-\lambda_i t}) \dots\dots\dots\dots\dots\dots\dots (5.11)$$

    b.   若 $\lambda_1 = \lambda_2 = \cdots\cdots = \lambda_n = \lambda$ 則

$$R(t) = 1 - (1 - e^{-\lambda t})^n \dots\dots\dots\dots\dots\dots\dots\dots (5.12)$$

$$\text{MTTF} = \frac{1}{\lambda} + \frac{1}{2\lambda} + \frac{1}{3\lambda} + \dots + \frac{1}{n\lambda} \dots\dots\dots\dots\dots (5.13)$$

(6) 以二項展開式評估可靠度 $(p+q)^n = \sum_{r=0}^{n} \binom{n}{r} p^{n-r} q^r$

設 $p = 1, q = (-R_i)$，則 $(1 - R_i)^n = \sum_{r=0}^{n} \binom{n}{r} (-1)^r (R_i)^r$

$\because \binom{n}{0} = 1$, $\therefore R(t) = 1 - (1 - R_i)^n = \sum_{r=1}^{n} \binom{n}{r} (-1)^{r-1} (R_i)^r$

若每一組件失效率可靠度均相同，則

$$R(t) = \sum_{r=1}^{n} \binom{n}{r} (-1)^{r-1} (e^{-r\lambda t}) \dots\dots\dots\dots\dots\dots (5.14)$$

$$\text{MTTF} = \sum_{r=1}^{n} \binom{n}{r} (-1)^{r-1} (\frac{1}{r\lambda}) = \int_0^{\infty} R(t) \, dt \dots\dots\dots\dots (5.15)$$

(7) 系統可靠度與並聯組件數的關係曲線

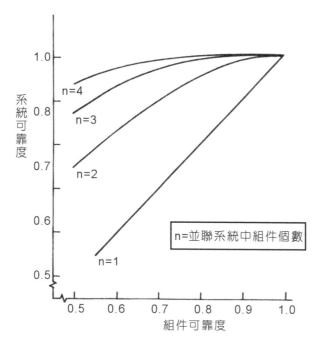

圖 5-3　系統可靠度與並聯組件數的關係曲線

例 5.5　某產品由三個組件並聯構成，其可靠度分別為 0.92，0.95，0.96，求該產品之可靠度？

解　$R_S = 1 - (1 - 0.92) \times (1 - 0.95) \times (1 - 0.96) = 1 - 0.00016 = 99.984\%$

例 5.6　某產品由三個組件並聯構成，每組件均有定值失效率 0.01 次/小時，求：

1. 每組件的 MTTF，
2. 每組件十小時內之可靠度，
3. 該產品十小時內之可靠度，
4. 該產品的 MTTF？

解

1. $\text{MTTF} = \dfrac{1}{0.01} = 100(\text{小時})$

2. $R(10) = e^{-0.01 \times 10} = 0.905$

3. $R(10) = 1 - (1 - e^{-0.01 \times 10})^3 = 1 - (1 - 0.905)^3 = 0.99914$

4. $\text{MTTF} = \dfrac{11}{6\lambda} = \dfrac{11}{6(0.01)} = 183.33(\text{小時})$

例 5.7 已知一有定值失效率系統的 MTTF，以及其終點壽命(End-of-life)可靠度為 0.9。求：

1. 以 MTTF 表示該系統的設計壽命，

2. 若將兩個該系統主動複聯，則複聯系統的設計壽命為原系統的幾倍？

解

1. $R = e^{-\lambda T} \Rightarrow T = \dfrac{1}{\lambda} \ln\left(\dfrac{1}{R}\right) = \ln\left(\dfrac{1}{R}\right) \times \text{MTTF}$

$$= \ln\left(\dfrac{1}{0.9}\right)\text{MTTF} = 0.105\,\text{MTTF}$$

2. 由(5.4)式：$R = 2e^{-\lambda T} - e^{-2\lambda T}$，設 $X = e^{-\lambda T}$，則

$X^2 - 2X + R = 0 \quad \Rightarrow \quad X = \dfrac{2 \pm \sqrt{4 - 4R}}{2} = 1 - \sqrt{1 - R}$

(因 $X \le 1$，故 "+" 號解不合)。

$e^{-\lambda T} = 1 - \sqrt{1 - R} \quad \Rightarrow \quad T = \ln\left(\dfrac{1}{1 - \sqrt{1 - R}}\right)$

$\text{MTTF} = \ln\left(\dfrac{1}{1 - \sqrt{1 - 0.9}}\right)\text{MTTF} = 0.380\,\text{MTTF}$

兩者相比：$\dfrac{0.380\,\text{MTTF}}{0.105\,\text{MTTF}} = 3.62$，複聯系統的設計壽命為原系統的 3.62 倍。

例 5.8　某組件可靠度為 0.7，求由 2, 3, 4 個該組件主動複聯之可靠度？

解　由(5.11)式，$R(2) = 1 - (1 - 0.7)^2 = 91\%$，

$R(3) = 1 - (1 - 0.7)^3 = 97.3\%$，$R(4) = 1 - (1 - 0.7)^4 = 99.2\%$

$\Rightarrow$　系統可靠度遞增，但邊際可靠度遞減。

例 5.9　欲將 MTTF 加倍，則須將幾個相同組件並聯？

解　由(5.15)式可得：$\dfrac{\text{MTTF}_n}{\text{MTTF}_1} = \displaystyle\sum_{r=1}^{n} (-1)^{r-1} \dfrac{n!}{r(n-r)!r!}$

$\Rightarrow$　$\dfrac{\text{MTTF}_2}{\text{MTTF}_1} = \dfrac{2!}{1(2-1)!1!} - \dfrac{2!}{2(2-2)!2!} = 1\dfrac{1}{2}$

$\dfrac{\text{MTTF}_3}{\text{MTTF}_1} = \dfrac{3!}{1(3-1)!1!} - \dfrac{3!}{2(3-2)!2!} + \dfrac{3!}{3(3-3)!3!} = 1\dfrac{5}{6}$

$\dfrac{\text{MTTF}_4}{\text{MTTF}_1} = \dfrac{4!}{1(4-1)!1!} - \dfrac{4!}{2(4-2)!2!} + \dfrac{4!}{3(4-3)!3!} - \dfrac{4!}{4(4-4)!4!} = 2\dfrac{1}{12}$

所以要至少 4 個相同組件並聯方可將 MTTF 加倍。

(另解：根據式(5.13)，n 個相同組件並聯後之

$\text{MTTF} = \dfrac{1}{\lambda} + \dfrac{1}{2\lambda} + \dfrac{1}{3\lambda} + \ldots + \dfrac{1}{n\lambda}$，

n=1，$\text{MTTF} = \dfrac{1}{\lambda}$

n=2，$\text{MTTF} = \dfrac{1}{\lambda} + \dfrac{1}{2\lambda} = \dfrac{3}{2}\dfrac{1}{\lambda}$

n=3，$\text{MTTF} = \dfrac{1}{\lambda} + \dfrac{1}{2\lambda} + \dfrac{1}{3\lambda} = \dfrac{11}{6}\dfrac{1}{\lambda}$

n=4，$\text{MTTF} = \dfrac{1}{\lambda} + \dfrac{1}{2\lambda} + \dfrac{1}{3\lambda} + \dfrac{1}{4\lambda} = 2\dfrac{1}{12}\dfrac{1}{\lambda}$

所以要至少 4 個相同組件並聯方可將 MTTF 加倍。)

3. 串並聯系統

(1) 越多零件串聯系統可靠度越低，若由多於 20 個零件串聯而成的系統，其各零件可靠度須高於 0.95 系統才會有用。

(2) 越多零件並聯系統可靠度越高，若由多於 10 個零件並聯而成的系統，再加入並聯的零件對系統可靠度幾乎毫無影響，此謂達到「回收遞減點(Point of diminishing return)」。

(3) 提昇零件可靠度，串聯和並聯系統的可靠度均可提昇。故欲提昇系統可靠度時，提昇零件可靠度應比複聯優先考慮。

例 5.10 某系統有 100 個零件，分成 3 個組件。組件 2(由 20 個可靠度為 0.93 的零件串聯而成)與組件 3(由 60 個可靠度為 0.96 的零件串聯而成)並聯後再與組件 1(由 20 個可靠度為 0.95 的零件串聯而成)串聯，求該系統之可靠度？

解

$R_1 = 0.95^{20} = 0.358$

$R_2 = 0.93^{20} = 0.234$

$R_3 = 0.96^{60} = 0.086$

$R = \left[1 - (1 - 0.234)(1 - 0.086)\right] \times 0.358 = 0.1074$

---

(4) 提升系統可靠度可有下列方法：

a. 減少零件數

b. 簡化系統結構

c. 提升零件可靠度

d. 將零件經過「燒入」程序

該如何選擇，須視設備的本質、成本及任務等因素而定。

(5)　將可靠度低的組件複聯

現有可靠度分別為 R_1 及 R_2 的 1、2 兩組件串聯，如圖 5-4(a)所示，設串聯後之系統可靠度為 R_a，則 $R_a = R_1 \times R_2$。此時應複聯可靠度較高或較低之組件才可得到較高之複聯系統可靠度？設複聯組件 1(如圖 5-4(b))與複聯組件 2(如圖 5-4(c))後之複聯系統可靠度分別為 R_b 及 R_c，則 $R_b = (2 - R_1)R_1R_2$，$R_c = R_1(2 - R_2)R_2$，兩者之差為 $R_b - R_c = R_1R(R_2 - R_1)$。所以若 $R_2 > R_1$ 則系統(b)較佳；若 $R_2 < R_1$ 則系統(c)較佳。故將可靠度較低的組件複聯，系統可得最大可靠度。

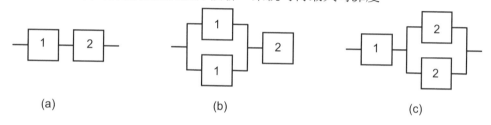

(a)　　　　　　　　　　(b)　　　　　　　　　　(c)

圖 5-4　將可靠度低或高的組件複聯比較

例 5.11　設 $R_1 = 0.7$，$R_2 = 0.95$。現欲將圖 5-4(b)系統之可靠度採增加一個組件之複聯方式提昇，問：應複聯 1 或複聯 2？

解

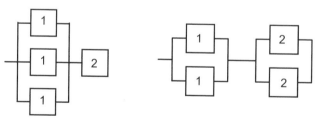

複聯 1 之系統可靠度 $= \left[1 - \left(1 - R_1\right)^3\right]R_2 = 0.973 \times 0.95 = 0.92435$

複聯 2 之系統可靠度 $= \left[1 - \left(1 - R_1\right)^2\right]\left[1 - \left(1 - R_2\right)^2\right]$

$$= 0.91 \times 0.9975 = 0.9077$$

⇒　當系統可靠度受一未複聯的低可靠度組件限制時，將較高可靠度組件複聯將是資源浪費。

(6) 高階複聯與低階複聯

    a. 高階複聯指將系統並聯(如圖 5-5(a))，低階複聯指組件並聯(如圖 5-5(b))。

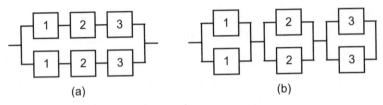

圖 5-5　高階複聯與低階複聯

高階複聯後的可靠度 $R_H = 1 - \left(1 - R_1 R_2 R_3\right)^2$

低階複聯後的可靠度 $R_L = \left(2R_1 - R_1^2\right)\left(2R_2 - R_2^2\right)\left(2R_3 - R_3^2\right)$

設 $R_1 = R_2 = R_3 = R$，則

$$R_H = 2R^3 - R^6 \text{，} R_L = \left(2R - R^2\right)^3$$

$$R_L - R_H = 6R^3 \left(1 - R\right)^2$$

    b. $\because \left(1 - R\right) > 0$，　$\therefore$ 低階複聯可得到較高的系統可靠度。

例 5.12 設 $R_1 = R_2 = R_3 = 0.7$。請分別求圖 5-5 中之系統採高階複聯與低階複聯之可靠度？

解　$R_H = 2\left(0.7\right)^3 - \left(0.7\right)^6 = 0.568531$，$R_L = \left[2\left(0.7\right) - \left(0.7\right)^2\right]^3 = 0.753571$

---

(7) 稀有事件近似值(Rare-event approximation)

    a. 因複聯系統係為得到很高的可靠度，也就是很低的失效率，故失效在複聯系統中極少發生，可視為稀有事件。

    b. 將可靠度以泰勒級數(Taylor series)展開：

$$R(t) = e^{-\lambda t} = \sum_{k=0}^{\infty} \frac{1}{k!}(-\lambda)^k t^k = 1 - \lambda t + \frac{1}{2}(\lambda t)^2 - \frac{1}{6}(\lambda t)^3 + \cdots\cdots$$

$\because \lambda t \ll 1$，　$\therefore$ 僅取前兩項，則

$$R(t) \cong 1 - \lambda t \quad\dots\dots\dots\dots\dots\dots\dots\dots\dots\dots\dots\dots (5.16)$$

c.　採取(5.16)式之近似值，可得

(a)　n 個組件串聯時之可靠度：$R(t) = \exp\left(-\sum_{i=1}^{n} \lambda_i t\right) \cong 1 - \sum_{i=1}^{n} \lambda_i t$

(b)　n 個組件並聯時之可靠度：$R(t) = 1 - \prod_{i=1}^{n}\left(1 - e^{-\lambda_i t}\right) \cong 1 - \prod_{i=1}^{n}\left(\lambda_i t\right)$

(c)　n 個相同組件串聯時之可靠度：$R(t) \cong 1 - n\lambda t$

(d)　n 個相同組件並聯時之可靠度：$R(t) \cong 1 - (\lambda t)^n$

例 5.13　系統結構如圖 5-6。設 $R_a = R_b = R$，$R_c = 1$，請以稀有事件近似值求系統之可靠度？

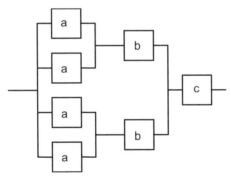

圖 5-6　例題 5.13 之系統方塊圖

解　將圖 5-6 合併如圖 5-7，則 $R_X = 2R - R^2$，

$R_Y = R_X \times R$，$R_Z = 2R_Y - R_Y^2$，　$R(t) = R_Z \times R_c$

$\Rightarrow$　$R(t) = \left(2R - R^2\right)R\left[2 - \left(2R - R^2\right)R\right] \times 1$

$= 4R^2 - 2R^3 - 4R^4 + 4R^5 - R^6$ ....(5.17)

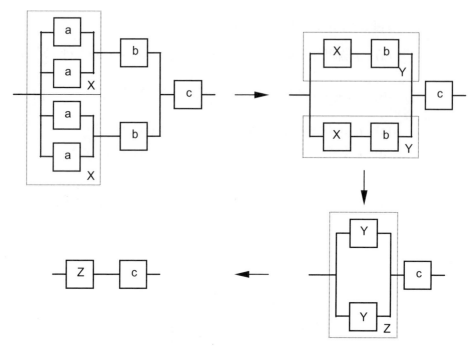

圖 5-7　圖 5-6 之合併圖

(1)　若取 $R(t) = e^{-\lambda t} \cong 1 - \lambda t$ ，代入(5.17)，得

$$R(t) = 4(1-\lambda t)^2 - 2(1-\lambda t)^3 - 4(1-\lambda t)^4 + 4(1-\lambda t)^5 - (1-\lambda t)^6$$

(2)　若改取 $R^n = e^{-n\lambda t} \cong 1 - n\lambda t + \frac{1}{2}n^2(\lambda t)^2$ ，代入(5.17)，得

$$R(t) = 4\left[1 - 2\lambda t + \frac{1}{2}2^2(\lambda t)^2\right] - 2\left[1 - 3\lambda t + \frac{1}{2}3^2(\lambda t)^2\right]$$

$$-4\left[1 - 4\lambda t + \frac{1}{2}4^2(\lambda t)^2\right] + 4\left[1 - 5\lambda t + \frac{1}{2}5^2(\lambda t)^2\right]$$

$$-\left[1 - 6\lambda t + \frac{1}{2}6^2(\lambda t)^2\right] = 1 - (\lambda t)^2$$

(3)　討論：

　　a.　設 $\lambda = 0.01$ 、 $t = 1$ ，代入(1)之結果得 $R(t) = 0.999898$ ，代入(2)之
　　　　結果得 $R(t) = 0.9999$ ；

若設 $\lambda = 0.001$、$t = 1$，代入(1)之結果得 $R(t) = 0.99999899$，代入(2)之結果得 $R(t) = 0.999999$。

b. 若方法(2)僅取前兩項，i.e. $e^{-n\lambda t} \cong 1 - n\lambda t$，結果如何？

(8) 共同失效模式(Common-mode failure)

a. 由共同原因引起數個組件或整個系統失效的事件模式稱為共同失效模式。此類系統通常係由一子系統串連一共同組件而成，如圖5-8所示。

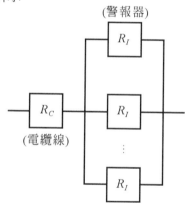

圖 5-8　共同失效模式示意圖

b. 設子系統由 n 個可靠度均為 $R_I = e^{-\lambda_I t}$ 的組件並聯而成，其中 λ_I 為每一組件之獨立失效率。則該子系統之可靠度為

$$R_{I \times n} = 1 - (1 - R_I)^n$$

c. 設該子系統與另一可靠度為 $R_C = e^{-\lambda_C t}$ 之組件（共同失效模式）串聯而成一系統，其中 λ_C 為共同模式失效率。則該串聯系統之可靠度為

$$R_n = R_C \times R_{I \times n} = R_C \left[1 - (1 - R_I)^n \right]$$

(a) 若 $n = 1$，則

$$R_1 = R_C \times R_{I \times 1} = e^{-\lambda_C t} \times e^{-\lambda_I t}$$

故 R_I 指的是組件本身的獨立可靠度，R_1 則是指共同模式加一個組件的總可靠度。

(b) 若 $n = 2$，則

$$R_2 = R_C \times R_{I \times 2} = R_C \left[1 - \left(1 - R_I \right)^n \right] = R_C \left[1 - \left(1 - R_I \right)^2 \right]$$

$$= e^{-\lambda_C t} \left[1 - \left(1 - e^{-\lambda_I t} \right)^2 \right]$$

(c) 若 $n = n$，則

$$R_n = R_C \times R_{I \times n} = R_C \left[1 - \left(1 - R_I \right)^n \right]$$

$$= e^{-\lambda_C t} \left[1 - \left(1 - e^{-\lambda_I t} \right)^n \right]$$

d. 設每一組件之總失效率均為 λ，因 λ 係在一 R_C 串聯一 R_I 之架構之下測試而得，故

$$\lambda = \lambda_I + \lambda_C$$

e. 設 $\beta = \dfrac{\lambda_C}{\lambda}$ 表 λ_C 在 λ 中所佔的比例，則 $\lambda_C = \beta\lambda$，$\lambda_I = (1 - \beta)\lambda$

可得

$$R_2 = e^{-\lambda_C t} \left[1 - \left(1 - e^{-\lambda_I t} \right)^2 \right] = \left[2 - e^{-(1-\beta)\lambda t} \right] e^{-\lambda t}$$

$$R_n = e^{-\lambda_C t} \left[1 - \left(1 - e^{-\lambda_I t} \right)^n \right] = e^{-\beta\lambda t} \left\{ 1 - \left[1 - e^{-(1-\beta)\lambda t} \right]^n \right\}$$

例 5.14 要求某型溫度感應器系統的設計壽命可靠度不低於 0.98，而已知一個感應器的可靠度只有 0.9。現工程師計劃將兩個感應器並聯，並聯後系統的可靠度應為 0.99，則可合乎要求。但當進行可靠度測試時，該並聯系統的可靠度卻只有 0.97。經研究後推知此結果係由共同失效模式所造成，故打算由下列二方法改善之：

1. 於並聯系統中再並聯一感應器，

2. 降低共同失效模式的失效率。

若感應器有定值失效率，求：

(1) 共同失效模式的可靠度？

(2) 該並聯系統代表共同失效模式的因子 β？

(3) 於並聯系統中再並聯一感應器是否能達到設計要求？

(4) 若欲使兩個感應器並聯的系統即可符合可靠度要求，則代表共同失效模式的因子 β 應降低多少%？

解　(1) 設電纜線之可靠度為 X，感應器之可靠度為 Y，則

$$\left.\begin{array}{l} R_1 = R_C \times R_I = 0.9 \\ R_2 = R_C \times \left[1 - \left(1 - R_I \right)^2 \right] = 0.97 \end{array}\right\} \Rightarrow \begin{array}{l} R_C = 0.9759036 \\ R_I = 0.9222222 \end{array},$$

共同失效模式的可靠度 $= 0.976$

(2) $R_1 = 0.9 = e^{-\lambda t} \Rightarrow \lambda t = -\ln(0.9) = 0.1053605$

$R_C = 0.9759036 = e^{-\lambda_C t} \Rightarrow \lambda_C t = -\ln(0.9759036) = 0.0243915$

$\beta = \dfrac{\lambda_C t}{\lambda t} = \dfrac{0.0243915}{0.1053605} = 0.2315$

(或利用公式 $R_2 = \left[2 - e^{-(1-\beta)\lambda t} \right] e^{-\lambda t} = \left[2 - e^{-(1-\beta)0.1053605} \right] 0.9 = 0.97$

$\Rightarrow \beta = 0.2315$)

(3) $R_3 = R_C \times \left[1 - \left(1 - R_I \right)^3 \right] = 0.9759036 \times \left[1 - \left(1 - 0.9222222 \right)^3 \right] = 0.975$ ，

故再並聯一感應器並不能達到設計要求。

(4) 改良後之 $\hat{R}_2 = \hat{R}_C \times \left[1 - \left(1 - 0.9222222 \right)^2 \right] = 0.98$

改良後之

$\hat{R}_C = 0.9859644 = e^{-\hat{\lambda}_C t} \Rightarrow \hat{\lambda}_C t = -\ln(0.9859644) = 0.0141350$

$R_I = e^{-\lambda_I t} = 0.9222222 \Rightarrow \lambda_I t = 0.0809691$

改良後之 $\hat{\beta} = \dfrac{\hat{\lambda}_C t}{\hat{\lambda} t} = \dfrac{\hat{\lambda}_C t}{\hat{\lambda}_C t + \lambda_I t} = \dfrac{0.0141350}{0.0141350 + 0.0809691} = 0.1486$

$\Rightarrow \Delta\beta\% = \dfrac{0.1486 - 0.2315}{0.2315} \times 100\% = -35.81\%$

(另解 1：改良後之

$\hat{R}_1 = \hat{R}_C \times R_I = 0.9859644 \times 0.9222222 = 0.9092783 = e^{-\hat{\lambda} t}$

$\Rightarrow \hat{\lambda} t = -\ln(0.9092783) = 0.0951041 \Rightarrow \hat{\beta} = \dfrac{\hat{\lambda}_C t}{\hat{\lambda} t} = \dfrac{0.0141350}{0.0951041} = 0.1486$

$$\Rightarrow \ \Delta\beta\% = \frac{0.1486 - 0.2315}{0.2315} \times 100\% = -35.81\%$$

另解 2：利用公式

$$\hat{R}_2 = \left[2 - e^{-(1-\hat{\beta})\hat{\lambda}t}\right]e^{-\hat{\lambda}t} = \left[2 - e^{-(1-\hat{\beta})0.0951041}\right]0.9092783 = 0.98$$

$$\Rightarrow \ \hat{\beta} = 0.1486 \ \Rightarrow \ \Delta\beta\% = \frac{0.1486 - 0.2315}{0.2315} \times 100\% = -35.81\% \ )$$

例 5.15 若某組件的設計壽命可靠度為 0.95，求：

1. 在無共同失效模式的前提下，兩組件並聯後的可靠度？
2. 若欲使(1)中之可靠度至少為 0.99，則共同失效模式的因子 β 至多為多少？

解 (1) $R_{(no\,common\,mode)} = \left[1 - (1 - 0.95)^2\right] = 0.9975$

(2) $R_1 = R_C \times R_I = e^{-\lambda_C t} \times e^{-\lambda_I t} = e^{-\lambda t} = 0.95 \ \Rightarrow \ \lambda t = -\ln 0.95 = 0.0512933$

利用公式 $R_2 = \left[2 - e^{-(1-\beta)\lambda t}\right]e^{-\lambda t} = \left[2 - e^{-(1-\beta)0.0512933}\right]0.95 = 0.99$

$$\Rightarrow \ \beta = 0.1613$$

---

4. 鏈式結構(Linked configuration)

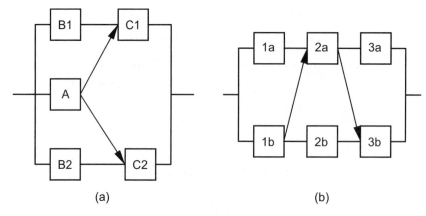

(a)   (b)

圖 5-9　鏈式結構系統之方塊圖

　　如圖 5-9 所示之例，無法用串並聯架構處理之系統。鏈式結構系統之可靠度評估有下列幾種方法可選擇。

(1)　眞値表法(Truth table method)

　　a.　設 R_A 爲組件 A 之可靠度，並以 1 表 A 可操作(其機率爲 R_A)以 0 表 A 失效(其機率爲 $1 - 1R_A = U_A$)，S 表系統可操作(其機率爲 R)以 F 表系統失效(其機率爲 $1 - R = U$)。

　　b.　將 n 個組件的所有狀態(即眞値表)均列出，共有 2^n 個狀態。

　　c.　依系統結構判定每個狀態爲 S 或 F。

　　d.　將所有 S 狀態的機率相加即爲系統可靠度。

例 5.16　請以眞値表法求圖 5-9(a)系統之可靠度？

設 $R_A = 0.3$，$R_{B1} = R_{B2} = 0.1$，$R_{C1} = R_{C2} = 0.2$

解

表 5-1　例題 5.16 之真値表

No	B1	B2	C1	C2	A	F or S	Probability
1	0	0	0	0	0	F	
2	0	0	0	0	1	F	
3	0	0	0	1	0	F	
4	0	0	0	1	1	S	0.03888
5	0	0	1	0	0	F	
6	0	0	1	0	1	S	0.03888
7	0	0	1	1	0	F	
8	0	0	1	1	1	S	0.00972
9	0	1	0	0	0	F	
10	0	1	0	0	1	F	
11	0	1	0	1	0	S	0.01008
12	0	1	0	1	1	S	0.00432
13	0	1	1	0	0	F	
14	0	1	1	0	1	S	0.00432
15	0	1	1	1	0	S	0.00252

表 5-1　例題 5.16 之真值表(續)

No	B1	B2	C1	C2	A	F or S	Probability
16	0	1	1	1	1	S	0.00108
17	1	0	0	0	0	F	
18	1	0	0	0	1	F	
19	1	0	0	1	0	F	
20	1	0	0	1	1	S	0.00432
21	1	0	1	0	0	S	0.01008
22	1	0	1	0	1	S	0.00432
23	1	0	1	1	0	S	0.00252
24	1	0	1	1	1	S	0.00108
25	1	1	0	0	0	F	
26	1	1	0	0	1	F	
27	1	1	0	1	0	S	0.00112
28	1	1	0	1	1	S	0.00048
29	1	1	1	0	0	S	0.00112
30	1	1	1	0	1	S	0.00048
31	1	1	1	1	0	S	0.00028
32	1	1	1	1	1	S	0.00012

將各個 S 狀態的機率依元件的失效與否計算出來，例如：

狀態 4：$S_4 = R_{C2} \times R_A \times U_{B1} \times U_{B2} \times U_{C1}$

$$= 0.2 \times 0.3 \times (1-0.1) \times (1-0.1) \times (1-0.2) = 0.03888$$

再將所有 S 狀態的機率加起來即為系統可操作的機率，亦即系統可靠度：
$R = \sum_i P_{S_i} = 0.13572$

(2)　分解法(Decomposition method)或稱「完全機率法(Total probability method)」或「貝氏定理法(Bayes rule method)」。

　　　a.　貝氏定理：若 X 事件發生的前提為 Y_1 或 Y_2 此二互斥事件之一發生，則 X 事件發生的機率為

　　　　　$P(X) = P\left(X|Y_1\right)P(Y_1) + P\left(X|Y_2\right)P(Y_2)$。

b. 將系統中之關鍵組件(有最多 Input 與 Output 之組件)設爲上述貝氏定理中之 Y，其可操作設爲 Y_1，失效設爲 Y_2。

c. 由系統結構分別求出當 Y_1 發生及 Y_2 發生時，系統爲 F 的機率 $P(X)$。

d. $1 - P(X)$ 即爲系統可靠度。(或是直接求當 Y_1 發生及 Y_2 發生時，系統爲 S 的機率)

e. (a) Bayes Rule

$$P(A \cap B) = P(B \cap A) \quad \Rightarrow \quad P(A|B)P(B) = P(B|A)P(A)$$

$$\Rightarrow \quad P(B|A) = \frac{P(A|B)P(B)}{P(A)}$$

(b) Total Probability Theorem

If B_1, B_2, $\cdots\cdots$, B_n, are mutually exclusive, collectively exhaustive events, the probability of another event A is

$$P(A) = P(A \cap B_1) + P(A \cap B_2) + \ldots + P(A \cap B_n)$$

$$= \sum_{i=1}^{n} P(A \cap B_i) = \sum_{i=1}^{n} P(A|B_i) \times P(B_i)$$

(c) Generalized Bayes Rule

Combining Bayes Rule and Total Probability Theorem yields

$$P(B_j|A) = \frac{P(A|B_j) \times P(B_j)}{\sum_{i=1}^{n} P(A|B_j) \times P(B_j)}$$

例 5.17　請以貝氏定理法求圖 5-9(a)系統之可靠度？

設 $R_A = 0.3$，$R_{B1} = R_{B2} = 0.1$，$R_{C1} = R_{C2} = 0.2$

解　1. 選 A 爲關鍵組件，故 Y_1 代表 A 可操作、Y_2 代表 A 失效。$P(X)$ 代表系統爲失效的機率。

2. Y_1 時須 C1 與 C2 同時失效系統才會失效，故

$$P(X|Y_1) = (1 - R_{C1})(1 - R_{C2})$$

3. Y_2 時須(B1 至 C1)與(B2 至 C2)同時失效系統才會失效，故

$$P(X|Y_2) = (U_{B1} \cup U_{C1}) \cap (U_{B2} \cup U_{C2}) = (1 - R_{B1}R_{C1})(1 - R_{B2}R_{C2})$$

4. $P(X) = P(X|Y_1)P(Y_1) + P(X|Y_2)P(Y_2)$

$$= [(1 - R_{C1})(1 - R_{C2})]R_A + [(1 - R_{B1}R_{C1})$$
$$(1 - R_{B2}R_{C2})](1 - R_A) = \ 系統失效機率$$

5. $R = 1 - P(X) = 1 - \left[(1 - 0.2)(1 - 0.2) \right]$

$$\times 0.3 + \left[(1 - 0.02)(1 - 0.02) \right] \times 0.7 = 0.13572$$

---

(3) 裴氏網路法(Petri net method)

裴氏網路(Petri nets)是一種通用型的數學工具，用以描述事件與條件間的關係。裴氏網路的基本符號包括：

○ ：位置(Place)，以一圓圈表示，代表事件

━━ ：立即變遷(Immediate transition)，以一細棒表示，代表事件轉移時無時間延遲

▬▬ ：時延變遷(Timed transition)，以一粗棒表示，代表事件轉移時有時間延遲

↑ ：弧(Arc)，以一箭頭表示，介於位置與變遷中間

● ：標記(Token)，以一圓點表示，含於位置中，代表資料

○̶ ：禁止弧(Inhibit arc)，以一圓圈與一直線表示，介於位置與變遷中間

如果輸入位置滿足了使能(Enable)條件，則變遷可發射(Fire)。變遷發射會使變遷之每一個輸入位置移出一個標記，並放入一個標記至其每一個輸出位置中。以裴氏網路表示事件間邏輯關係的基本結構示於圖5-10，其中 P、Q、R 分別代表不同的事件。圖中變遷之輸入位置可分為既定型和條件型兩類。既定型只有一個輸出弧，而條件型之輸出弧則有多個。在既定型位置中之標記僅有一個去處，也就是說如果該位置擁有一個標記，在變遷發射後，將使得該變遷之每一輸出位置均

獲得一個標記。然而在條件型中的標記因有多個出路，故會依條件將系統導引至不同的狀況。以圖 5-10 中「TRANSFER OR」的裴氏網路為例，是 Q 或是 R 位置會由 P 位置取得一個標記，將視例如：機率、外界動作、或內部狀況等條件而定。另依耗時情形可將變遷分為三類：轉移過程沒有任何時間延遲的變遷稱為立即變遷，而延遲時間為常數者稱為時延變遷。第三類稱為隨機變遷(Stochastic transistion)，是用來模擬時間延遲為隨機的程序。裴氏網路對於系統建模(System modeling)失效分析及預防維護甚為有用。

Logic relation	TRANSFER	AND	OR	TRANSFER AND	TRANSFER OR	INHIBITION
Description	If P then Q	If P AND Q then R	If P OR Q then R	If P then Q AND R	If P then Q OR R	If P AND Q' then R
Boolean function	Q=P	R=P*Q	R=P+Q	Q=R=P	Q+R=P	R=P*Q'
Petri nets	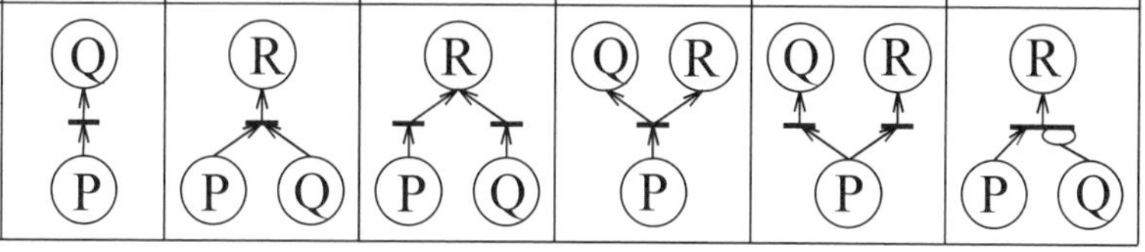 Q P	R P Q	R P Q	Q R P	Q R P	R P Q

圖 5-10　以裴氏網路表示事件間邏輯關係的基本結構

以裴式網路評估系統可靠度的方法以及與其他常用方法的比較，請參考附錄 A。

5.　m-out-of-n(m/n)系統

m/n 系統指在 n 個單元中須有 m 個單元可操作系統方可操作的系統。

(1)　各單元之可靠度均相等

　　a.　設 $R_1 = R_2 = \cdots = R_n = R$，且失效機率 $p = 1 - R$，失效個數為 N，則 n 個單元中有 i 個單元失效的機率為

$$P(N = i) = \binom{n}{i} p^i (1 - p)^{n-i} \text{。}$$

　　b.　m/n 系統可操作指失效單元數不得多於$(n - m)$個，其機率即為其可靠度

$$R_{m/n} = P(N \le n-m) = \sum_{i=0}^{n-m} \binom{n}{i} p^i (1-p)^{n-i}$$

$$= 1 - \sum_{i=n-m+1}^{n} \binom{n}{i} p^i (1-p)^{n-i} \text{。} \dots\dots\dots\dots\dots\dots\dots\dots\dots (5.18)$$

c. 其失效的機率為

$$p_{m/n} = P(N > n-m) = \sum_{i=n-m+1}^{n} \binom{n}{i} p^i (1-p)^{n-i}$$

$$= 1 - \sum_{i=0}^{n-m} \binom{n}{i} p^i (1-p)^{n-i} \text{。}$$

d. 若失效機率很低(或說可靠度很高)，i.e. $p \ll 1$，則 $(1-p)^{n-i} \approx 1$，可靠度可改寫成：$R_{m/n} \approx \sum_{i=0}^{n-m} \binom{n}{i} p^i \approx 1 - \sum_{i=n-m+1}^{n} \binom{n}{i} p^i$。然而當次冪 i 增大時，$p^i \approx 0$，故可靠度可再改寫成：

$$R_{m/n} \approx 1 - \binom{n}{n-m+1} p^{n-m+1} \text{。}$$

e. 其失效的機率亦可再改寫為：$p_{m/n} \approx \binom{n}{n-m+1} p^{n-m+1}$。

設失效率為定值 λ 則 $p = 1 - R = 1 - e^{\lambda t}$，若為稀有事件則 $p = 1 - R = 1 - e^{\lambda t} = 1 - (1 - \lambda t) = \lambda t$，失效的機率可再改寫為：

$$p_{m/n} \approx \binom{n}{n-m+1} (\lambda t)^{n-m+1} \text{。}$$

(2) m-out-of-n failed 系統：指在 n 個單元中若有 m 個單元失效則系統方失效的系統。

(3) Consecutive m-out-of-n 系統：指在 n 個單元中須有 m 個單元可操作(失效)系統方可操作(失效)的系統，但此 m 個單元有接連順序的排列關係(Consecutively ordered)。

例 5.18 某通訊系統有四個波道，其中任三個波道可操作則系統可操作，求該系統之可靠度。

解　設 x_1、x_2、x_3、及 x_4 分別代表該四個波道可操作的事件，則系統可靠度為

$$R = \left(x_1 \cap x_2 \cap x_3\right) \cup \left(x_1 \cap x_2 \cap x_4\right) \cup \left(x_1 \cap x_3 \cap x_4\right)$$
$$\cup \left(x_2 \cap x_3 \cap x_4\right) \cup \left(x_1 \cap x_2 \cap x_3 \cap x_4\right)$$
$$= P\left(x_1 x_2 x_3 + x_1 x_2 x_4 + x_1 x_3 x_4 + x_2 x_3 x_4\right)$$
$$= P\left(x_1 x_2 x_3\right) + P\left(x_1 x_2 x_4\right) + P\left(x_1 x_3 x_4\right) + P\left(x_2 x_3 x_4\right)$$
$$- 6P\left(x_1 x_2 x_3 x_4\right) + 4P\left(x_1 x_2 x_3 x_4\right) - P\left(x_1 x_2 x_3 x_4\right)$$
$$= P\left(x_1 x_2 x_3\right) + P\left(x_1 x_2 x_4\right) + P\left(x_1 x_3 x_4\right) + P\left(x_2 x_3 x_4\right) - 3P\left(x_1 x_2 x_3 x_4\right)$$

若每個波道的可靠度均相等為 $q(q = 1 - p)$ 則 $R = 4q^3 - 3q^4$

另解：

由(5.18)式：$R_{3/4} = P(N \le 4 - 3) = \sum_{i=0}^{4-3} \binom{4}{i} p^i \left(1 - p\right)^{4-i}$

$$= 1 - \sum_{i=4-3+1}^{4} \binom{4}{i} p^i \left(1 - p\right)^{4-i} \text{ 或寫成}$$

$$R_{3/4} = \sum_{i=3}^{4} \binom{4}{i} q^i \left(1 - q\right)^{4-i} = \binom{4}{3} q^3 \left(1 - q\right)^1 + \binom{4}{4} q^4 \left(1 - q\right)^0 = 4q^3 - 3q^4$$

6.　安全失效與危險失效

a.　依系統之作動時機正確與否可將失效分成危險失效與安全失效兩種。

b.　以汽車安全氣囊(Airbag)為例，危險發生該爆發而未爆發之失效稱「危險失效(Fail-to-danger)」，反之危險未發生不該爆發而爆發之失效稱「安全失效(Fail-safe)」。

c.　複聯方式通常係為降低危險失效，但安全失效機率會隨複聯數增加而上升。

d.　在 1/n(1-out-of-n)並聯系統中以 p_d 及 p_s 分別代表危險失效及安全失效發生的機率則：

該系統發生危險失效的機率：$p_{(1/n),d} = \left(p_d\right)^n$

該系統發生安全失效的機率：$p_{(1/n),s} = 1 - \left(1 - p_s\right)^n$

(n 個元件中任一發生即發生)

e. 在 m/n 並聯系統中

該系統發生危險失效的機率：

$$p_{(m/n),d} = P(N > n-m) = \sum_{i=n-m+1}^{n} \binom{n}{i} p_d^{\ i} \left(1-p_d\right)^{n-i} \approx \binom{n}{n-m+1} p_d^{\ n-m+1}$$

該系統發生安全失效的機率：

$$p_{(m/n),s} = P(N \geq m) = \sum_{i=m}^{n} \binom{n}{i} p_s^{\ i} \left(1-p_s\right)^{n-i} \approx \binom{n}{m} p_s^{\ m}$$

f. m 固定時，n 愈大系統發生危險失效的機率愈低，但系統發生安全失效的機率愈高。但 n 太大時共同失效模式會大幅限制危險失效機率的改善。反之，n 固定時，m 愈大系統發生危險失效的機率愈高，但系統發生安全失效的機率愈低。

例 5.19 某可靠度工程師受命設計一 m/n 系統，已知每一組件之危險失效及安全失效機率分別爲 $p_d = 10^{-2}$，$p_s = 10^{-2}$。爲降低成本 n 必須爲最小，且要求系統發生危險失效的機率須低於 10^{-4}，系統發生安全失效的機率須低於 10^{-2}。求 n 及 m 值。

解

m/n	該系統發生安全失效的機率 $p_{(m/n),s} \approx \binom{n}{m} p_s^{\ m}$	該系統發生危險失效的機率 $p_{(m/n),d} \approx \binom{n}{n-m+1} p_d^{\ n-m+1}$
1/1	10^{-2}	10^{-2}
1/2	2×10^{-2}	10^{-4}
2/2	10^{-4}	2×10^{-2}
1/3	3×10^{-2}	10^{-6}
2/3	3×10^{-4}	3×10^{-4}
3/3	10^{-6}	3×10^{-2}
1/4	4×10^{-2}	10^{-8}
◎ 2/4	6×10^{-4}	4×10^{-6}
3/4	4×10^{-6}	6×10^{-6}
4/4	10^{-8}	4×10^{-2}

$n = 4$，$m = 2$，即 2/4 系統可滿足要求。

■ 5.3　可靠度配當(Reliability allocation)

　　一產品的可靠度確定後，該如何分配到組成該產品之各分項，以達成系統可靠度的要求稱為可靠度配當。

1.　考慮因素

　　(1)　產品架構之複雜性

　　(2)　組件之重要性或關鍵性

　　(3)　產品之安全性

　　(4)　使用環境條件

　　(5)　維修便易性(進手)

　　(6)　成本

　　(7)　產品未來發展性

　　(8)　工程水準

2.　常用配當法

　　(1)　等量配當法

　　　　a.　將各分項之可靠度均視為相同的配當法，適用於對各分項可靠度之影響因素尚未了解時所作的初步配當。

　　　　b.　$R_S = R_1 \times R_2 \times \cdots\cdots \times R_n = \left(R_i \right)^n$　$\Rightarrow$　$R_i = \sqrt[n]{R_S}$

　　　　　　其中 R_S：系統可靠度，

　　　　　　　　R_i：第 i 分項可靠度配當值。

例 5.20　設某產品由五個分項串聯而成，若該產品的可靠度目標為 0.95，請以等量配當法求各分項之可靠度。

解　$R_i = \sqrt[n]{R_S} = \sqrt[5]{0.95} = 0.9898$

(2) ARINC 配當法

a. 由美國航空無線電公司(Aeronautical Radio Incorporation, ARINC)
於 1958 年所發表的方法。

b. 假設：

(a) 各分項為串聯

(b) 各分項之任務時間與系統之任務時間相同

(c) 失效率為常數

$$R_i = \left(R_S\right)^{W_i} , \qquad W_i = \frac{\lambda_i}{\sum\limits_{i=1}^{n} \lambda_i}$$

W_i：第 i 分項之權重(Weighting factor)

λ_i：第 i 分項之失效率

例 5.21 設某產品由五個分項串聯而成，若該產品的可靠度目標為 0.998，請
以 ARINC 法求各分項之可靠度。設任務時間 t = 1000h，且 $\lambda_1 = 0.0001$
次/h，$\lambda_2 = 0.0002$ 次/h，$\lambda_3 = 0.0003$ 次/h，$\lambda_4 = 0.0004$ 次/h，
$\lambda_5 = 0.0005$ 次/h。

解

$\overline{\lambda}_{1,1000h} = 0.0001 \times 1000 = 0.1$(次/h)，

$\overline{\lambda}_{2,1000h} = 0.0002 \times 1000 = 0.2$(次/h)，

$\overline{\lambda}_{3,1000h} = 0.0003 \times 1000 = 0.3$(次/h)，

$\overline{\lambda}_{4,1000h} = 0.0004 \times 1000 = 0.4$(次/h)，

$\overline{\lambda}_{5,1000h} = 0.0005 \times 1000 = 0.5$(次/h)，

$\sum\limits_{i=1}^{5} \overline{\lambda}_{i,1000h} = 0.1 + 0.2 + 0.3 + 0.4 + 0.5 = 1.5$ (次/h)。

$W_1 = \dfrac{0.1}{1.5} = 0.0667 \quad \Rightarrow \quad R_1 = \left(0.998\right)^{0.0667} = 0.99987$

$W_2 = \dfrac{0.2}{1.5} = 0.1333 \quad \Rightarrow \quad R_2 = \left(0.998\right)^{0.1333} = 0.99973$

$$W_3 = \frac{0.3}{1.5} = 0.2000 \quad \Rightarrow \quad R_3 = (0.998)^{0.2000} = 0.99960$$

$$W_4 = \frac{0.4}{1.5} = 0.2667 \quad \Rightarrow \quad R_4 = (0.998)^{0.2667} = 0.99947$$

$$W_5 = \frac{0.5}{1.5} = 0.3333 \quad \Rightarrow \quad R_5 = (0.998)^{0.3333} = 0.99933$$

(3) AGREE 配當法

　由美國 AGREE 於 1958 年所發表的方法。

　考慮分項之複雜性(零件數)及關鍵性(重要性)。

$$R_i(t_i) = e^{-\lambda_i \times t_i} \quad , \quad \lambda_i = \frac{n_i \left[-\ln(R_S) \right]}{N \times C_i \times t_i}$$

　其中 R_S ：系統可靠度

　　　　N ：零件總數

　　　　n_i ：第 i 分項之零件數

　　　　C_i ：第 i 分項之關鍵因子

　　　　t_i ：第 i 分項之任務時間

　　　　R_i ：第 i 分項之可靠度

例 5.22　某產品有四個分項，要求連續使用 10 小時的可靠度目標為 0.95，請以 AGREE 法求各分項之可靠度。設第一及第三分項為關鍵分項，第二分項任務時間為 9 小時，關鍵因子為 0.95；第四分項任務時間為 8 小時，關鍵因子為 0.9。各分項之零件數：$n_1 = 30$，$n_2 = 100$，$n_3 = 50$，$n_4 = 80$。

解　　$N = 30 + 100 + 50 + 80 = 260$

$$\lambda_1 = \frac{30 \left[-\ln(0.95) \right]}{260 \times 1 \times 10} = 5.918 \times 10^{-4} \text{(次//小時)} , \quad R_1(10) = e^{-\lambda_1 \times 10} = 0.9941$$

$$\lambda_2 = \frac{100 \left[-\ln(0.95) \right]}{260 \times 0.95 \times 9} = 23.074 \times 10^{-4} \text{(次//小時)} , \quad R_2(9) = e^{-\lambda_2 \times 9} = 0.97945$$

$$\lambda_3 = \frac{50\left[-\ln(0.95)\right]}{260 \times 1 \times 10} = 9.864 \times 10^{-4} (\text{次/小時}), \qquad R_3(10) = e^{-\lambda_3 \times 10} = 0.99018$$

$$\lambda_4 = \frac{80\left[-\ln(0.95)\right]}{260 \times 0.9 \times 8} = 21.92 \times 10^{-4} (\text{次/小時}), \qquad R_4(8) = e^{-\lambda_4 \times 8} = 0.98262$$

(4) 評點(Rating)配當法

    a. 由多位有經驗人員給予每個分項 1~10 的評點，再整合所有人員評點的結果，給予該分項一個配當評點。

    b. 平均評點：

$$可靠度：R_i = 1 - (1 - R_S) \times \left\{ \frac{\prod\limits_{j=1}^{M}\left[\dfrac{\sum\limits_{k=1}^{N} Z_{ijk}}{N}\right]}{\sum\limits_{i=1}^{L}\left[\prod\limits_{j=1}^{M}\left(\dfrac{\sum\limits_{k=1}^{N} Z_{ijk}}{N}\right)\right]} \right\}$$

其中 N：參與評定人員數

    L：分項數

    M：每分項的影響因子數

    Z_{ijk}：第 k 位專家對第 i 分項的第 j 個影響因子所給的評點數

    R_i：第 i 分項的可靠度

    R_S：系統的可靠度

$$失效率：\lambda_i = \left\{ \frac{\sum\limits_{j=1}^{M}\left[\dfrac{\sum\limits_{k=1}^{N} Z_{ijk}}{N}\right]}{\sum\limits_{i=1}^{L}\left[\prod\limits_{j=1}^{M}\left(\dfrac{\sum\limits_{k=1}^{N} Z_{ijk}}{N}\right)\right]} \right\} \times \lambda_S$$

c.　乘法評定評點：

$$W_i = \frac{\prod\limits_{j=1}^{m} Z_{ij}}{\sum\limits_{i=1}^{n}\prod\limits_{j=1}^{m} Z_{ij}} \quad , \quad \lambda_i = W_i \times \lambda_s$$

d.　加法評定評點：

$$W_i = \frac{\sum\limits_{j=1}^{m} Z_{ij}}{\sum\limits_{i=1}^{n}\sum\limits_{j=1}^{m} Z_{ij}} \quad , \quad \lambda_i = W_i \times \lambda_s$$

3.　串並聯系統可得最大可靠度之元件配置法

設一串並聯系統由 n 個次系統串聯而成，且每一次系統由 m 個元件並聯而成，如圖 5-11 所示。

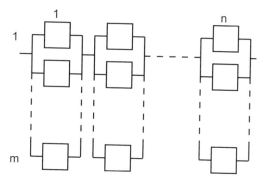

圖 5-11　串並聯系統的方塊圖

設

K_i：第 i 個次系統，$i = 1, 2, …, n$

$u = n \times m$，代表該系統之元件總數

C_j：第 j 個元件，$j = 1, 2, …, u$

P_j：C_j 可操作的機率(i.e. C_j 的可靠度)

則該系統之可靠度 R 可表示為 $\qquad R = \prod_{i=1}^{n}\left(1 - \prod_{j \in K_i}\left(1 - P_j\right)\right)$

由串並聯架構之特性可知:盡量使每一次系統之可靠度相等則可使該系統得最大可靠度。Baxter 與 Harche(1992)提出一由上而下(Top Down Heuristic, TDH)的元件配置法,其步驟如下:

a. 將所有元件按可靠度由高到低排列,亦即 $P_1 \geq P_2 \geq \cdots\cdots \geq P_u$

b. 將可靠度最高之前 n 個元件(C_j,$j = 1, 2, \cdots\cdots, n$)依序配置給該 n 個次系統(K_i,$i = 1, 2, \cdots\cdots, n$);亦即配置每個次系統之第一個元件。

c. 將 C_j 配置給 K_{2n+1-j},$j = n + 1, \cdots\cdots, 2n$;亦即將剩餘的元件依可靠度由高至低從 K_n,K_{n-1},$\ldots$,K_1 往回依序配置每個次系統之第二列的元件。

d. 如果 $m > 2$ 則每個次系統之第三列以下的元件繼續依以下步驟配置;若 $m = 2$ 則停止。

e. 設 $v = 2$

f. 評估 $R_i^{(v)} = \left(1 - \prod_{j \in K_i}\left(1 - P_j\right)\right)$,$i = 1, 2, \cdots\cdots, n$,

($R_i^{(v)}$ 代表組件數為 v 之 K_i 的可靠度)

若 $R_i^{(v)}$ 是第 j 小(jth smallest, $j = 1, 2, \cdots\cdots, n$),則將 C_{vn+i} 配置給 K_i。

g. 設 $v = v + 1$。

h. 如果 $v < m$ 則重複步驟 f 及步驟 g;若 $v = m$ 則停止。

例 5.23 某產品有六個阻值相同且可互換的電阻,可靠度分別為 0.95,0.75,0.85,0.65,0.40,0.55。因空間的限制,該六個電阻必須(3,2)排列,亦即兩組串聯且每組有三電阻並聯。

1. 請以 TDH 法配當此系統,

2. 請以 BUH(Bottom-up heuristic)法配當並與(1)之結果比較。

解　步驟 1：將元件依可靠度排列如下：0.95，0.85，0.75，0.65，0.55，0.40

步驟 2：將 C_1 配置給 K_1，C_2 配置給 K_2。

步驟 3：將 C_3 配置給 K_2，C_4 配置給 K_1。

步驟 4：設 $v = 2$。

步驟 5：分別計算 K_1 及 K_2 次系統的可靠度：

$$R_1^{(2)} = 1 - (1 - 0.95)(1 - 0.65) = 0.9825$$

$$R_2^{(2)} = 1 - (1 - 0.85)(1 - 0.75) = 0.9625$$

因為 $R_2^{(2)} < R_1^{(2)}$，所以將 C_5 配置給 K_2，將 C_6 配置給 K_1。

步驟 6：$v = 3 = m$，所以停止。

配置的結果為

次系統 K_1	次系統 K_2
C_1	C_2
C_4	C_3
C_6	C_5

可靠度 $R_{\text{TDH}} = \left[1 - (1 - 0.95)(1 - 0.65)(1 - 0.40) \right]$
$$\left[1 - (1 - 0.85)(1 - 0.75)(1 - 0.55) \right] = 0.972802$$

若改採 BUH 之方法(i.e.所有元件按可靠度由低到高排列)則配置的結果為

次系統 K_1	次系統 K_2
C_6	C_5
C_3	C_4
C_2	C_1

可靠度 $R_{\text{BUH}} = \left[1 - (1 - 0.40)(1 - 0.75)(1 - 0.85) \right]$
$$\left[1 - (1 - 0.55)(1 - 0.65)(1 - 0.95) \right] = 0.9698$$

通常 R_{TDH} 會高於 R_{BUH}。

■ 5.4 動態評估模式(Dynamic evaluation model)

動態模式是指對「狀態會隨時間而變化的系統」進行可靠度評估的模式。

1. 馬可夫過程(Markov process)

    (1) 馬可夫狀態(State)：系統的組件呈操作或失效的組合情形。

    (2) 一系統有 n 個組件則有 2^n 個馬可夫狀態。

    (3) 設一系統有 a、b、c 三組件，則該系統有 8 個馬可夫狀態如表 5-2。

表 5-2　三組件之 8 個馬可夫狀態表

組　件	狀　態　編　號							
	1	2	3	4	5	6	7	8
a	O	F	O	O	F	F	O	F
b	O	O	F	O	F	O	F	F
c	O	O	O	F	O	F	F	F

O：Operational，F：Failed

系統於某狀態下是否失效須視元件間組合的架構而定。若該系統為串聯，則於表 5-2 中之狀態#2~#8 該系統均為失效；若該系統為並聯，則僅狀態#8 時該系統為失效；若為 a、b 並聯後再與 c 串聯，則狀態#4~#8 該系統均為失效。

    (4) 設系統於時間 t 時為狀態 i 的機率為 $P_i(t)$，則系統之可靠度為
    $R(t) = \sum_{i \in O} P_i(t)$ ，亦即將所有可操作狀態之機率相加。或
    $R(t) = 1 - \sum_{i \in F} P_i(t)$ 。

    (5) $P_1(0) = 1$，且 $P_i(0) = 0$　$\forall\, i \neq 1$(起始狀態一定是狀態#1)；
    $\sum_i P_i(t) = 1$ (任一時間下，系統僅能有一個狀態)。

2.　二獨立組件之可靠度分析

(1)　二獨立組件共有四個馬可夫狀態(如圖 5-12(a)所示)，狀態間的轉移如圖 5-12(b)所示。其中狀態#1 只出不進，稱爲吐出狀態(Spitting state)；而狀態#4 只進不出，稱爲吸收狀態(Absorbing state)。

組件	狀　態　編　號			
	1	2	3	4
a	O	F	O	F
b	O	O	F	F

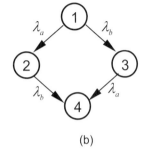

(a)　　　　　　　　　　　　　　　　(b)

圖 5-12　二獨立組件的四個馬可夫狀態及狀態轉移圖

(2)　$\because \lambda(t) = \dfrac{P(T < t + \Delta t \,|\, T > t)}{\Delta t}$ 爲在 $T > t$ 之前提下，將失效發生在 $T < t + \Delta t$ 的機率平均分配在時間 Δt 內所得的失效率，$\therefore \lambda(t)\Delta t = P(T < t + \Delta t \,|\, T > t)$ 爲在 $T > t$ 之前提下，失效發生在 $T < t + \Delta t$ 的機率。故 $\lambda_a \Delta t P_1(t)$ 爲在 Δt 時間內系統由狀態#1 轉移至狀態#2 的轉移率(Transition rate)。同理，在 Δt 時間內系統由狀態#1 轉移至狀態#3 的轉移率爲 $\lambda_b \Delta t P_1(t)$。現定義移出之機率爲負、移入之機率爲正，故於狀態#1 之狀態移轉機率可表示爲

$$P_1(t + \Delta t) - P_1(t) = -\lambda_a \Delta t P_1(t) - \lambda_b \Delta t P_1(t)$$

或

$$\frac{d}{dt} P_1(t) = -\lambda_a P_1(t) - \lambda_b P_1(t) \quad\text{..} (5.19)$$

上式即稱爲狀態#1 之狀態移轉方程式(State transition equation)。

同理可得

$$\frac{d}{dt}P_2(t) = \lambda_a P_1(t) - \lambda_b P_2(t) \quad\text{.............}\quad (5.20)$$

$$\frac{d}{dt}P_3(t) = \lambda_b P_1(t) - \lambda_a P_3(t) \quad\text{.............}\quad (5.21)$$

$$\frac{d}{dt}P_4(t) = \lambda_b P_2(t) + \lambda_a P_3(t) \quad\text{.............}\quad (5.22)$$

(3) 由(5.19)式可解得狀態#1 之機率爲

$$P_1(t) = e^{-(\lambda_a + \lambda_b)t} \quad\text{.............}\quad (5.23)$$

將(5.23)式代入(5.20)式得

$$\frac{d}{dt}P_2(t) + \lambda_b P_2(t) = \lambda_a e^{-(\lambda_a + \lambda_b)t}$$

解得狀態#2 之機率爲

$$P_2(t) = e^{-(\lambda_b)t} - e^{-(\lambda_a + \lambda_b)t} \quad\text{.............}\quad (5.24)$$

同理狀態#3 之機率爲

$$P_3(t) = e^{-(\lambda_a)t} - e^{-(\lambda_a + \lambda_b)t} \quad\text{.............}\quad (5.25)$$

因此，狀態#4 之機率爲

$$P_4(t) = 1 - \sum_{i=1}^{3} P_i(t) = 1 - e^{-(\lambda_a)t} - e^{-(\lambda_b)t} + e^{-(\lambda_a + \lambda_b)t} \quad\text{.............}\quad (5.26)$$

(4) 若該二組件爲串聯，則 $R(t) = \sum_{i \in O} P_i(t) = P_1(t) = e^{-(\lambda_a + \lambda_b)t}$

若該二組件爲並聯，則 $R(t) = 1 - \sum_{i \in F} P_i(t) = 1 - P_4(t)$

$$= e^{-(\lambda_a)t} + e^{-(\lambda_b)t} - e^{-(\lambda_a + \lambda_b)t}$$

3. 被動複聯系統的可靠度分析

(1) 不考慮切換裝置的失效

假設切換裝置不會失效，僅只考慮主要組件(Primary unit)a 及備用組件(Backup unit)b 會失效的理想狀況。此時之馬可夫狀態及狀態轉移如圖 5-13 所示。

組件	狀 態 編 號			
	1	2	3	4
a	O	F	O	F
b	O	O	F	F

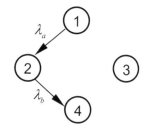

圖 5-13　不考慮切換裝置失效的馬可夫狀態及狀態轉移圖

a. 假設備用組件於備用狀態不會失效，亦即備用組件不會在主要組件失效之前發生失效，故 $P_3(t) = 0$。

b. 三個狀態轉移方程式如下：

$$\frac{d}{dt}P_1(t) = -\lambda_a P_1(t)$$

$$\frac{d}{dt}P_2(t) = \lambda_a P_1(t) - \lambda_b P_2(t)$$

$$\frac{d}{dt}P_4(t) = \lambda_b P_2(t)$$

解得各狀態之機率為

$$P_1(t) = e^{-(\lambda_a)t}$$

$$P_2(t) = \frac{\lambda_a}{\lambda_b - \lambda_a}[e^{-(\lambda_a)t} - e^{-(\lambda_b)t}]$$

$$P_4(t) = 1 - \frac{1}{\lambda_b - \lambda_a}[\lambda_b e^{-(\lambda_a)t} - \lambda_a e^{-(\lambda_b)t}]$$

故該系統之可靠度為

$$R(t) = \sum_{i \in O} P_i(t) = P_1(t) + P_2(t) = e^{-(\lambda_a)t} + \frac{\lambda_a}{\lambda_b - \lambda_a}[e^{-(\lambda_a)t} - e^{-(\lambda_b)t}]$$

$$= \frac{1}{\lambda_b - \lambda_a}[\lambda_b e^{-(\lambda_a)t} - \lambda_a e^{-(\lambda_b)t}] \quad\quad\quad (5.27)$$

c.　$\lambda_a = \lambda_b$ 時之可靠度及 MTTF

將(5.27)式改寫成

$$R(t) = e^{-(\lambda_a)t} + \frac{\lambda_a}{\lambda_b - \lambda_a} e^{-(\lambda_a)t}[1 - e^{-(\lambda_b - \lambda_a)t}] \quad\quad (5.28)$$

而　$e^{-(\lambda_b - \lambda_a)t} = 1 - (\lambda_b - \lambda_a)t + \frac{1}{2}(\lambda_b - \lambda_a)^2 t^2 - \cdots \quad (5.29)$

將(5.29)式代入(5.28)式得

$$R(t) = e^{-(\lambda_a)t} + \lambda_a e^{-(\lambda_a)t}\left[t - \frac{1}{2}(\lambda_b - \lambda_a)t^2 + \cdots\cdots\right]$$

若 $\lambda_a = \lambda_b = \lambda$，則

$$R(t) = e^{-(\lambda)t} + \lambda e^{-(\lambda)t}[t] = (1 + \lambda t)e^{-\lambda t} \quad\quad\quad (5.30)$$

$$\text{MTTF} = \int_0^\infty R(t)dt = \int_0^\infty (1 + \lambda t)e^{-\lambda t}dt = \frac{2}{\lambda} = \frac{4}{2\lambda}$$

(主動複聯之 MTTF $= \frac{2}{\lambda} - \frac{1}{2\lambda} = \frac{3}{2\lambda}$，可見被動複聯之可靠度高於主動

複聯可靠度)

例 5.24 設某產品之 MTTF 為 3000 小時，現將執行連續 500 小時的任務。
求：

1. 任務期間之可靠度，
2. 若將兩個該產品以被動複聯方式連結，求連結後該系統之
   MTTF 及任務期間之可靠度？

解

1. $R(t) = \exp(-\dfrac{1}{3000} \times 500) = 0.846$

2. $\text{MTTF} = \dfrac{2}{\lambda} = 2 \times 3000 = 6000$（小時）

     由(5.30)式，

$$R(t) = (1 + \lambda t)e^{-\lambda t} = \left(1 + \dfrac{1}{3000} \times 500\right)\exp\left(-\dfrac{1}{3000} \times 500\right) = 0.988$$

(2) 考慮切換裝置的失效

     設切換裝置之(需求)失效率為 p，此時之馬可夫狀態及狀態轉移圖如圖 5-14 所示。

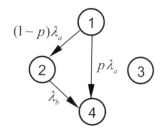

圖 5-14　考慮切換裝置失效的馬可夫狀態及狀態轉移圖

a.   假設備用組件於備用狀態不會失效，亦即備用組件不會在主要組件失效之前發生失效，故 $P_3(t) = 0$。

b.   三個狀態轉移方程式如下：

$$\dfrac{d}{dt}P_1(t) = -(1-p)\lambda_a P_1(t) - p\lambda_a P_1(t) = -\lambda_a P_1(t)$$

$$\dfrac{d}{dt}P_2(t) = (1-p)\lambda_a P_1(t) - \lambda_b P_2(t) \quad , \quad \dfrac{d}{dt}P_4(t) = \lambda_b P_2(t) + p\lambda_a P_1(t)$$

解得各狀態之機率為

$$P_1(t) = e^{-(\lambda_a)t} \quad , \quad P_2(t) = (1-p)\dfrac{\lambda_a}{\lambda_b - \lambda_a}[e^{-(\lambda_a)t} - e^{-(\lambda_b)t}]$$

故該系統之可靠度為

$$R(t) = \sum_{i \in O} P_i(t) = P_1(t) + P_2(t) = e^{-(\lambda_a)t} + \frac{(1-p)\lambda_a}{\lambda_b - \lambda_a}[e^{-(\lambda_a)t} - e^{-(\lambda_b)t}]$$

若 $\lambda_a = \lambda_b = \lambda$，則

$$R(t) = [1 + (1-p)\lambda t]e^{-\lambda t} \quad\text{.................................... (5.31)}$$

由(5.31)式可知：

(a) p 遞增，備用組件之價值遞減。

(b) $p = 1$ 時，備用組件已無法提昇系統可靠度。

例 5.25 設某產品之可靠度為 0.9，且備用組件於備用狀態時之失效率可忽略不計，但切換開關之失效率必須考慮。現欲採兩組件複聯方式提高可靠度，求：

1. 主動複聯與被動複聯之可靠度各為多少？

2. 欲使被動複聯之可靠度高於主動複聯，則切換開關之失效率至多為多少？

3. 若任務時間非常短，欲使被動複聯之可靠度高於主動複聯，則切換開關之失效率之需求為何？

解

1. $R(t) = \exp(-\lambda T) = 0.9 \quad\Rightarrow\quad \lambda T = 0.1054$

    主動複聯可靠度 $R_A = 2e^{-\lambda T} - e^{-2\lambda T} = 0.98999$

    (比較：$R_A = [1 - (1 - 0.9)(1 - 0.9)] = 0.99$)

    被動複聯可靠度 $R_P = [1 + (1 - p)\lambda t]e^{-\lambda t} = 0.9948 - 0.094856p$

2. 令 $0.98999 = 0.9948 - 0.094856p$，解得 $p = 0.0507$

3. 若任務時間非常短，由稀有事件之結果

    主動複聯可靠度 $R_A \cong 1 - (\lambda T)^2$

    被動複聯可靠度 $R_P = [1 + (1 - p)\lambda t]e^{-\lambda t} \cong [1 + (1 - p)\lambda t]$

$$(1 - \lambda T) = 1 - p\lambda T - \left(\frac{1}{2} - p\right)(\lambda T)^2$$

令 $R_A = R_P$，得 $p = \dfrac{\dfrac{1}{2}\lambda T}{1 - \lambda T} \approx \dfrac{1}{2}\lambda T$。

所以任務時間 T 愈短，則 p 須跟著愈小，否則無法使被動複聯之可靠度高於主動複聯。

4. 裴氏網路法(Petrinet method)

以裴氏網路進行系統建模以及可靠度評估的方法請參考附錄 B、C。裴氏網路是一紙上作業的工具，但裴氏網路可以實現成特用晶片(Application-Specific Integrated Circuit，ASIC)，以硬體形式執行其功能。裴氏網路實現成特用晶片的相關研討請參考附錄 D。

❖ 習 題

1. 某產品經測試得其可靠度爲 0.9，

    (1) 若在相同條件下，兩個該產品並聯所得之可靠度爲多少？

    (2) 若上小題之情況經測試得可靠度爲 0.97，求：

        (a) 共同模式失效率？

        (b) 該產品之獨立失效率？

        (c) 共同模式失效因子 β 之值？

    (3) 若要求兩個該產品並聯之後之可靠度爲 0.98，則共同模式失效因子 β 之值應爲多少？

2. 請寫出

    (1) Bayes' rule，

    (2) Total probability theorem，

    (3) Generalized Bayes's rule。

3. 某系統由 6 個裝置以鍊式結構組成如圖，各裝置之可靠度如下：$R_A = 0.9$，$R_B = 0.8$，$R_C = 0.7$，$R_D = 0.6$，$R_E = 0.5$，$R_F = 0.4$。請以

    (1) Decomposition method，

    (2) Petri net/Matrix method，求該系統之不可靠度。

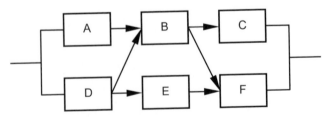

4. 某產品有四分項，要求連續使用 10 小時之 R = 0.95，其中第一及第三分項爲關鍵分項，第二分項之任務時間爲 8 小時，關鍵因子爲 0.9，第四分

項之任務時間爲 7 小時，關鍵因子爲 0.8，各分項零件數爲 $N_1 = 20$，$N_2 = 100$，$N_3 = 60$，$N_4 = 80$。請以 AGREE 法求各分項之可靠度配當。(答案請標示至小數點以下第四位)

5. 某系統有三個零件並聯組成，每一零件之失效率 $\lambda = 0.01$ 次/小時，求：

(1) 該系統之可靠度，

(2) 該系統 10 小時內仍可運作之機率？

(3) 該系統之 MTTF？

6. 某系統由三個可靠度均爲 0.9 之元件串聯而成。求：

(1) 該系統之可靠度？

(2) 若採高階複聯，複聯後之可靠度？

(3) 若採低階複聯，複聯後之可靠度？

7. 某水庫洩洪閘門之啓動系統的需求失效機率爲 0.0001，然不該啓動而啓動之機率爲 0.00001。爲確保水庫水位過高時能及時洩洪，現將三組啓動系統採高階主動複聯，求：

(1) 該複聯系統之危險失效機率及安全失效機率各爲多少？

(2) 若該複聯系統係 2-out-of-3 系統，其危險失效機率及安全失效機率各爲多少？

8. 某裝置之失效率爲 0.001 次/hr，現以一失效率爲 0.002 次/hr 之裝置與其被動複聯。假設 Backup unit 於備用狀態下不會失效，請以 Markov Process 求於 100h 之內：

(1) 該系統處於失效狀態的機率？

(2) 若考慮切換裝置之失效率爲 0.001，求該系統之可靠度。(答案需計算至小數點以下第 7 位，四捨五入至第 6 位)

9. Demand pacemaker 的方塊圖如下，請將其轉換為 Petri net 以描述其動作。

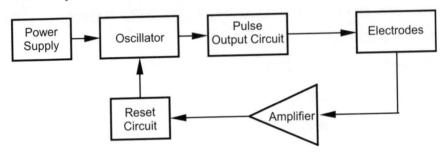

Chapter 6

可靠度試驗

■ 6.1 緒言

可靠度試驗(Reliability test)是指「針對產品進行可靠度估計和驗證的測試活動」。相關用詞：

1. 完全數據(Complete data)：試驗活動延續至所有受測物品均失效才停止之過程中所紀錄之數據。

2. 非參數法(Non-parametric method)：將試驗所得之數據直接點繪，不必與某一特定分布適配的數據分析方法。

3. 截略測試(Censoring test)：在所有受測物品均失效之前即停止試驗之測試方法，亦即該測試之數據為不完全(Incomplete)數據。

4. 加速壽命測試(Accelerated life test)：以失效率較正常操作為高的操作方法或環境加諸受測物品，藉以加速終結受測物品之壽命，以達到提早完成測試工作目的的試驗方法。

5. 可靠度成長(Reliability growing)：在產品量產上市前(即雛型階段)，以失效分析的方法掌握該產品失效的原因，據此修改設計進而使量產之該產品的可靠度較雛型時之可靠度為高。

■ 6.2 可靠度試驗的規劃

產品在設計階段已經各種失效分析方法掌握該產品預期的失效狀況，但也許會有出乎意料的失效模式發生，故可靠度試驗的方案須盡量涵蓋各種狀況，如儲存、處理、運送、修復等。

1. 變異性的考量

   (1) 產品變異的主因來自於由設計至實現為硬體所涉及的製造流程(例如同色環但不同阻值的電阻器)。

   (2) 因產品變異性的緣故，受測物品不能只有一個。

   (3) 決定受測品個數應考慮的項目：

      a. 主要變數能被試驗掌握的程度

      b. 失效的嚴重性

      c. 費用

   (4) 除非常複雜、昂貴、產量非常少等因素之產品外，受測品個數不應少於四個。

2. 時間效應

   (1) 有瞬間失效率隨時間遞增現象的產品，如機械組件會受磨耗、疲乏、腐蝕等因素而失效者，應著重耐久測試。

   (2) 無瞬間失效率隨時間遞增現象的產品，如半導體元件，其失效通常有80%~90%發生於操作之初的 200 小時之內，故應著重測試個數。

   (3) 可修復產品之可靠度會隨修復次數和年限而逐漸降低，故維修保養亦應包含於測試計劃之之中。

■ 6.3 可靠度試驗的內容

1. 試驗內容

    (1) 不希望發生失效的測試

        a. 功能試驗(Functional test)

        b. 環境試驗(Environmental test)

    (2) 故意造成失效的測試

        a. 壽命測試(Life test)

        b. 安全性試驗(Safety test)

2. 試驗分類

    可靠度試驗的分類如圖 6-1 所示。

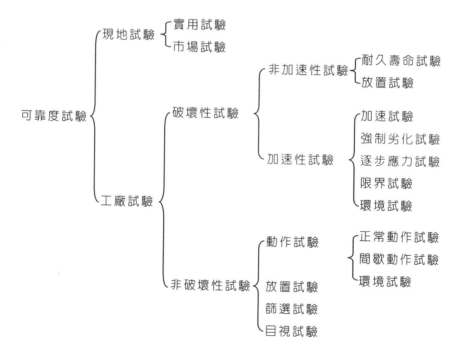

圖 6-1　可靠度試驗的分類

3. 試驗順序與方法

可靠度試驗的順序與方法可歸納如圖 6-2 所示。

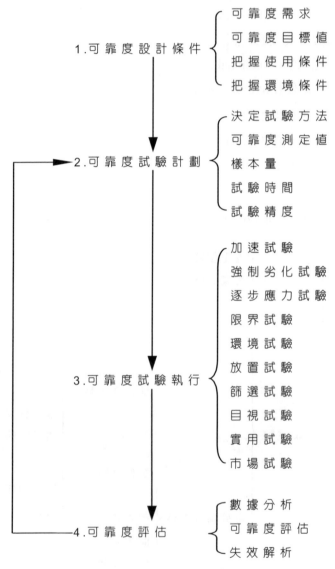

圖 6-2　可靠度試驗的順序與方法

4. 通過及不通過二項抽樣測試(Go-no-go test)

　　將所有受測品均操作一固定時間，然後求出於此時間中失效數佔總測試數之比例。此類測試有助於推測該產品之可靠度，但對於時間相依的可靠度則無法提供資訊。

5. 環境測試

　　(1) 一般常見會影響可靠度的環境因子計有：

　　　　a.　溫度(Temperature)

　　　　b.　振動(Vibration)

　　　　c.　衝擊(Shock)

　　　　d.　溼度(Humidity)

　　　　e.　電力變化(Power variation)

　　　　f.　灰塵與鹽霧(Dirt & salt spray)

　　　　g.　人員(People)

　　　　h.　電磁效應(Electro-magnetic effect)

　　　　i.　靜電(Static electricity)

　　　　j.　輻射(Radiation)

　　　　k.　黴菌(Fungus)

　　　　l.　噪音(Noise)

　　(2) 大多數試驗均爲單一環境測試(Single Environment Test, SET)，對某些產品亦須進行「組合環境可靠度測試(Combined Environmental Reliability Test, CERT)」，圖 6-3 即爲一針對運輸工具之電子系統的環境測試週期圖。

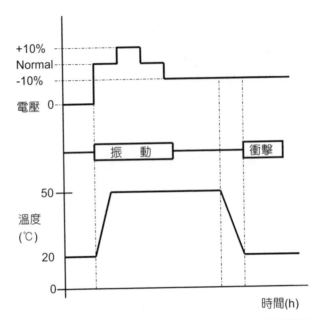

圖 6-3　針對運輸工具之電子系統的環境測試週期圖

(3) 以下數點是決定環境測試條件時須注意事項

    a. 環境因素變化率：不僅考慮因素的極大與極小值，亦須考慮因素的變化率(Rate of change)，例如溫度急劇變化所造成的影響。

    b. 操作與閒置的環境差異：例如潮濕性鏽蝕對閒置物品的影響遠大於操作中之物品。

    c. 組合環境所造成的影響：通常遠大於其中任一單一環境因素的影響。

    d. 振動與衝擊的方向與型態。

    e. 不同狀況下的環境差異：諸如操作、儲存、運輸、維修時的環境差異性。

■ 6.4 　非參數法(Non-parametric method)

　　非參數法係指將試驗所得之數據直接點繪,不必與某一特定分布適配的數據分析方法。其數據可分為「未分組數據」與「分組數據」兩類。

1. 未分組數據(Ungrouped data):或稱個別數據,得自可靠度試驗。例如表 6-1。

表 6-1　未分組數據

i	0	1	2	3	4	5	6	7	8	9
t_i	0.00	0.41	0.58	0.75	0.83	1.00	1.08	1.17	1.25	1.35

(1) 有 $n + 1$ 個失效時間 t_i ($i = 0, 1, 2,\cdots\cdots, n$)稱為「順序統計值(Rank statistics)」。其中 n 為受測品個數,i 為失效個數。

(2) 在 t_i 時之可靠度

$$\hat{R}(t_i) = \frac{x}{n} = \frac{n-i}{n} = 1 - \frac{i}{n} \quad\text{..(6.1)}$$

其中 x 稱為「倖存單元數(Survival units)」。t_i 時之可靠度所對應之累積失效分配函數為 $\quad \hat{F}(t_i) = 1 - \hat{R}(t_i) = \frac{i}{n}$

(3) (6.1)式有缺陷(可由例 6.1 中看出,當 $i = 9$ 時 $R(t_i) = 0$),將其修正為

$$\hat{F}(t_i) = \frac{i}{n+1} \quad\Rightarrow\quad \hat{R}(t_i) = \frac{n+1-i}{n+1} = 1 - \frac{i}{n+1} \quad\text{.................................(6.2)}$$

(4) $f(t_i) = -\dfrac{d}{dt}R(t_i)$ 改寫成差分形式: $\hat{f}(t_i) = -\dfrac{\hat{R}(t_{i+1}) - \hat{R}(t_i)}{t_{i+1} - t_i}$ ....(6.3)

例 6.1 請利用表 6-1 之數據建構該產品之可靠度、失效機率密度函數、失效率的圖形。

解

i	t_i	$t_{i+1} - t_i$	$R(t_i)$ (6.2)	$R(t_i)$ (6.1)	$f(t_i)$ (6.2)	$f(t_i)$ (6.1)	$\lambda(t_i)$ (6.2)	$\lambda(t_i)$ (6.1)
0	0	0.41	1.0	1.0	0.244	0.270	0.244	0.270
1	0.41	0.17	0.9	0.889	0.588	0.635	0.653	0.714
2	0.58	0.17	0.8	0.778	0.588	0.653	0.735	0.839
3	0.75	0.08	0.7	0.667	1.25	1.388	1.786	2.080
4	0.83	0.17	0.6	0.556	0.588	0.659	0.980	1.185
5	1.00	0.08	0.5	0.444	1.25	1.388	2.500	3.126
6	1.08	0.09	0.4	0.333	1.11	1.233	2.775	3.703
7	1.17	0.08	0.3	0.222	1.25	1.388	4.167	6.252
8	1.25	0.10	0.2	0.111	1.00	1.110	5.000	10
9	1.35	-	0.1	0	-	-	-	∞

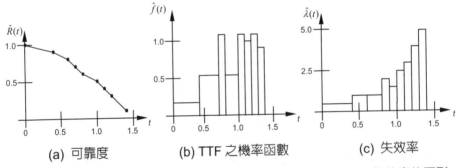

(a) 可靠度　　　　　(b) TTF 之機率函數　　　　　(c) 失效率

圖 6-4　根據表 6-1 之數據建構之可靠度、失效機率密度函數、失效率的圖形

2. 分組數據(Grouped data)：得自現場經驗。僅有一時段內之失效數據，而沒有個別確實失效時間之數據，例如表 6-2。

表 6-2　分組數據

時段	$0 \le t < 3$	$3 \le t < 6$	$6 \le t < 9$	$9 \le t < 12$	$12 \le t < 15$	$15 \le t < 18$
失效個數	21	10	7	9	2	1

(1) 試驗記錄 n 個受測品於 m 個時段 $t_i (i = 1, 2, \cdots\cdots, m)$ 中仍未失效之個數為 $n_i (i = 1, 2, \cdots\cdots, m)$ 個。

(2) 在 t_i 時之可靠度

$$\hat{R}(t_i) = \frac{n_i}{n} \quad\cdots\cdots\cdots\cdots\cdots\cdots\cdots\cdots\cdots\cdots\cdots\cdots\cdots\cdots\cdots\cdots\cdots\cdots (6.4)$$

(3) 因各 t_i 間之失效記錄為不連續，$t_i \sim t_{i+1}$ 間之可靠度可以內插法(Linear interpolation)求得近似值。

(4) 將(6.4)式代入(6.3)式可得

$$\hat{f}(t_i) = -\frac{\hat{R}(t_{i+1}) - \hat{R}(t_i)}{t_{i+1} - t_i} = -\frac{\dfrac{n_{i+1}}{n} - \dfrac{n_i}{n}}{t_{i+1} - t_i} = \frac{n_i - n_{i+1}}{(t_{i+1} - t_i)n} \quad\cdots\cdots\cdots (6.5)$$

(5) 若不考慮連續性，則

$$\hat{\lambda}(t) = \frac{\hat{f}(t)}{\hat{R}(t)} = \frac{n_i - n_{i+1}}{(t_{i+1} - t_i)n_i} \quad\cdots\cdots\cdots\cdots\cdots\cdots\cdots\cdots\cdots\cdots\cdots (6.6)$$

(6) 若改將(6.2)式代入(6.3)式可得

$$\hat{f}(t_i) = -\frac{\hat{R}(t_{i+1}) - \hat{R}(t_i)}{t_{i+1} - t_i} = -\frac{\dfrac{n+1-(i+1)}{n+1} - \dfrac{n+1-i}{n+1}}{t_{i+1} - t_i} = \frac{1}{(t_{i+1} - t_i)(n+1)} \quad\cdots (6.7)$$

(7) 因各 t_i 間之失效記錄為不連續，故 $\lambda(t)$ 改由下式求得

$$\lambda(t) = \frac{P(T < t + \Delta t \mid T > t)}{\Delta t} \quad\Rightarrow\quad \lambda(t)\Delta t = P(T < t + \Delta t \mid T > t)$$

$$= \frac{P\big[(T < t + \Delta t) \cap (T > t)\big]}{P(T > t)} = \frac{P(t < T < t + \Delta t)}{P(T > t)} = \frac{f(t)\Delta t}{R(t)}$$

現設 $t = t_i$ 以及 $\Delta t = (t_{i+1} - t_i)$，則

$$\lambda(t_i)(t_{i+1} - t_i) = \frac{f(t_i)(t_{i+1} - t_i)}{R(t_i)} \quad\cdots\cdots\cdots\cdots\cdots\cdots\cdots\cdots\cdots\cdots (6.8)$$

將(6.2)式及(6.7)式代入(6.8)式可得

$$\hat{\lambda}(t_i) = \frac{1}{(t_{i+1} - t_i)(n+1-i)}$$ .... (6.9)

例 6.2 請利用表 6-2 之數據建構該產品之可靠度、失效機率密度函數、失效率的圖形。

解

i	t_i	n_i	$R(t_i)$ (6.4)	$R(t_i)$ (6.2)	$f(t_i)$ (6.5)	$f(t_i)$ (6.7)	$\lambda(t_i)$ (6.6)	$\lambda(t_i)$ (6.9)
0	0	50	1.00	1.00	0.140	0.0065	0.140	0.0065
1	3	29	0.58	0.980	0.067	0.0065	0.115	0.0067
2	6	19	0.38	0.961	0.046	0.0065	0.123	0.0068
3	9	12	0.24	0.941	0.060	0.0065	0.250	0.0069
4	12	3	0.06	0.922	0.013	0.0065	0.222	0.0071
5	15	1	0.02	0.902	0.0067	0.0065	0.333	0.0072
6	18	0	0.00	0.882	-	0.0065	-	0.0074

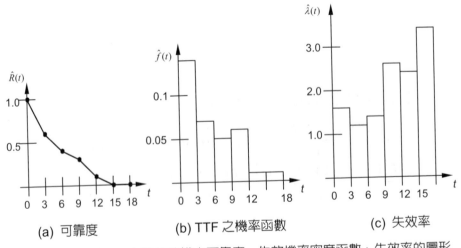

(a) 可靠度   (b) TTF 之機率函數   (c) 失效率

圖 6-5 根據表 6-2 之數據建構之可靠度、失效機率密度函數、失效率的圖形

(8) 設 f_i 為區間 (t_{i+1}, t_i) 中之數值，$\bar{t_i} = \dfrac{1}{2}(t_i + t_{i+1})$ ，區間寬度 $\Delta_i = t_{i+1} - t_i$ 則

平均數 $\hat{\mu} = \displaystyle\sum_{i=0}^{m-1} \bar{t_i} f_i \Delta_i$ ，變異數 $\hat{\sigma}^2 = \displaystyle\sum_{i=0}^{m-1} \bar{t_i}^2 f_i \Delta_i - \hat{\mu}^2$

■ 6.5　截略與加速

1. 單次截略(Single censored)

   (1) 指因部分受測品於所有受測品均失效前被挪用，或因其他原因而終止之測試，又稱為「自右截略(Censored on the right)」。

   (2) 可分為

       a. Type I(定時型)：執行至事先決定之時間即終止的測試。

       b. Type II(定數型)：執行至事先決定之失效個數發生時即終止的測試。

   (3) 對非參數法而言，又可視為「失效截略(Failure-censored)」。亦即有 n 個受測品，但只能得到前 $i(i < n)$ 個失效的數據。

   (4) 因測試所得為前數個失效的數據，屬失效率曲線中之早夭期的部分，故不適用於估計整個壽命的平均數或變異數，但對設定保固期卻相當有用。

   (5) 截略的個數或時間可設定在實驗者關切的地方，如保固期或設計壽命。

2. 多次截略(Multiple censored)

   (1) 指因部分受測品於測試的各個時間內因造成失效的機制並非在研究的範圍之內、或挪作他用、或其他原因而被剔除的實驗。例如車輛傳動軸的壽命測試，會將因車禍撞毀的車輛自實驗數據中剔除。

(2) 由(6.2)式可得

$$\hat{R}(t_{i-1}) = \frac{n+1-(i-1)}{n+1} = \frac{n+2-i}{n+1}$$ .... (6.10)

由(6.2)式及(6.10)式可得

$$\frac{\hat{R}(t_i)}{\hat{R}(t_{i-1})} = \frac{n+1-i}{n+2-i}$$

所以

$$\hat{R}(t_i) = \frac{n+1-i}{n+2-i}\hat{R}(t_{i-1})$$ .... (6.11)

i.e. 受測品在 t_i 不失效的機率 ＝ 由(t_{i-1} 至 t_i)不失效的機率 × (在 t_{i-1} 不失效的機率)

(3) 然若受測品於 t_i 時仍未失效而執行截略(該數據會加一個 " + " 號)，則 $\hat{R}(t_i) = \hat{R}(t_{i-1})$，故多次截略的可靠度為

$$\hat{R}(t_i) = \left(\frac{n+1-i}{n+2-i}\right)^{\delta_i} \hat{R}(t_{i-1})，\text{ where } \delta_i = \begin{cases} 1 & failure @ t_i \\ 0 & functioning @ t_i \end{cases}$$

或是

$$R(0) = R(t_0) = 1，\quad \hat{R}(t_i) = \prod_{k=0}^{i} \hat{R}(t_k) = \prod_{k=0}^{i} \left(\frac{n+1-k}{n+2-k}\right)^{\delta_k}$$ .... (6.12)

(4) (6.12)式係用來計算失效時(t_i)之可靠度，兩點間的可靠度可以線性內插估計。

例 6.3 請利用下表之數據計算各 t_i 之可靠度，並繪出可靠度曲線。

1	27	6	85+
2	39	7	93
3	40+	8	102
4	54	9	135+
5	69	10	144

解

i	t_i	$\dfrac{n+1-i}{n+2-i}$ (n = 10)	$\hat{R}(t_i)$
1	27	10/11	$R(27) = (10/11) \times 1.0 = 0.9090$
2	39	9/10	$R(39) = (9/10) \times 0.9090 = 0.8181$
3	40+	8/9	$R(40) = 0.8181$
4	54	7/8	$R(54) = (7/8) \times 0.8181 = 0.7159$
5	69	6/7	$R(69) = (6/7) \times 0.7159 = 0.6136$
6	85+	5/6	$R(85) = 0.6136$
7	93	4/5	$R(93) = (4/5) \times 0.6136 = 0.4909$
8	102	3/4	$R(102) = (3/4) \times 0.4909 = 0.3681$
9	135+	2/3	$R(135) = 0.3681$
10	144	1/2	$R(144) = (1/2) \times 0.3681 = 0.18409$

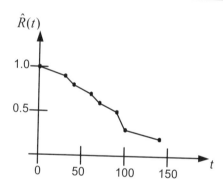

圖 6-6　例 6.3 之可靠度曲線

3. 逐次試驗(Sequential test)

   (1) 試驗時數不事先決定，乃是由試驗結果查詢逐次試驗計劃圖來判斷允收或拒收。

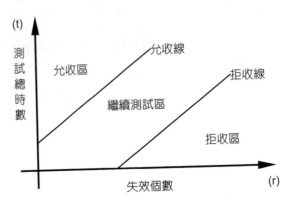

圖 6-7　逐次試驗計劃圖

(2) 允收線方程式：$t_1 = \dfrac{\ln\left(\dfrac{\theta_0}{\theta_1}\right)}{\left(\dfrac{1}{\theta_1} - \dfrac{1}{\theta_0}\right)} \times r - \dfrac{\ln\left(\dfrac{\beta}{1-\alpha}\right)}{\left(\dfrac{1}{\theta_1} - \dfrac{1}{\theta_0}\right)}$ .... (6.13)

拒收線方程式：$t_2 = \dfrac{\ln\left(\dfrac{\theta_0}{\theta_1}\right)}{\left(\dfrac{1}{\theta_1} - \dfrac{1}{\theta_0}\right)} \times r - \dfrac{\ln\left(\dfrac{1-\beta}{\alpha}\right)}{\left(\dfrac{1}{\theta_1} - \dfrac{1}{\theta_0}\right)}$ .... (6.14)

其中：θ_0：指定之 MTBF

θ_1：消費者願意接受最短之 MTBF

α：生產者風險

β：消費者風險

r：受測品失效個數

t：測試總時數

例 **6.4**　某產品之 $\theta_0 = 480$ 小時，$\theta_1 = 160$ 小時，$\alpha = 0.1$，$\beta = 0.1$。現隨機取樣 10 個進行逐次試驗，測試進行中若有受測品失效則立即更換新品替代，直至 10 個受測品受測至實驗終止。該測試之前 5 個失效時間記錄如下：

1	20.3 小時
2	50.9 小時
3	99.8 小時
4	110.2 小時
5	148.9 小時

請問：　1.　根據在第 5 個產品失效時之數據，是否可允許該批產品出廠？

　　　　2.　若數據仍落在繼續測試區，則應再繼續測試多久且不發生失效才可允許出廠？

解　將指定參數代入(6.13)及(6.14)式則分別可得

允收線方程式：$t_1 = \dfrac{1.099}{\left(\dfrac{1}{160} - \dfrac{1}{480}\right)} \times r + \dfrac{2.197}{\left(\dfrac{1}{160} - \dfrac{1}{480}\right)} = 263.7 \times r + 527.28$

拒收線方程式：$t_2 = \dfrac{1.099}{\left(\dfrac{1}{160} - \dfrac{1}{480}\right)} \times r - \dfrac{2.197}{\left(\dfrac{1}{160} - \dfrac{1}{480}\right)} = 263.7 \times r - 527.28$

根據失效記錄可求得受測總時數如下表

失效序號	失效時間(小時)	受測總時數(小時)
1	20.3	$20.3 \times 10 = 203$
2	50.9	$50.9 \times 10 = 509$
3	99.8	$99.8 \times 10 = 998$
4	130.2	$130.2 \times 10 = 1302$
5	148.9	$148.9 \times 10 = 1489$

將允收線、拒收線、各次失效發生時之受測總時數繪成逐次試驗計劃圖如下。

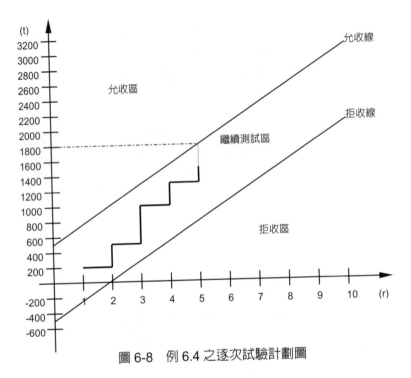

圖 6-8　例 6.4 之逐次試驗計劃圖

(1)　由逐次試驗計劃圖可知，在第 5 個產品失效時(即 1489 小時)，仍落在繼續測試區，無法判定是否允收，故不可允許該批產品出廠。

(2)　將 $r = 5$ 代入允收線方程式

$$t_1 = 263.7 \times 5 + 523.28 = 1842 \ (小時)$$

故應再繼續測試

$(1842 \div 10) - 148.9 = 35.3$ (小時)

且不發生失效才可允許出廠。

(3)　若測試曲線先與拒收線相交，則可判定為不可出廠。

4.　加速壽命測試

(1)　以失效率較正常操作為高的操作方法或環境施以受測物品，藉以提早完成測試工作的試驗方法。但無法處理即使是在停用狀態也會衰退而引發失效的系統之壽命測試。

(2) 縮時測試(Compressed-time test)

    a.   係指故意增加產品操作的次數(或時間)，以加速該產品失效的測試方法。但於測試時，產品的負載及環境仍與正常使用時相同。屬於「耐久測試」。

    b.   最適用於失效率與操作週期次數(Number of cycle)有關，而非關於連續長時間操作的產品。例如汽車的車門鎖，正常使用時大約一天開關數次，但於縮時測試時可增加為每分鐘數次。另亦適用於持續與間歇交互操作的系統，例如汽車引擎，正常使用時大約一天數小時，但於縮時測試時可使之連續運轉至失效為止。

    c.   應合理縮時以避免因縮時而產生於正常操作時沒有的失效機制。例如車門開關頻率過高，以致因散熱不及造成超溫損壞。

    d.   縮時測試無法發現須長時間才能引發的失效機制，例如生鏽。

(3) 增壓測試(Advanced-stress test)

    a.   係指故意將產品置於高應力水準之下，以加速該產品失效的測試方法。由高應力水準之下所測得的數據再推算該產品於正常應力水準之下的可靠度。屬於「耐用測試」。

    b.   環境應力諸如：溫度、機械壓力、電壓、輻射密度……等等，施加產品之應力可為單一或數種應力之組合。

    c.   可由數個不同高應力水準之下所測得的數據，以數據適配(Fit)的方法推算該產品於正常應力水準之下的可靠度，如圖 6-9 所示。

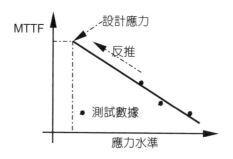

圖 6-9　以數據適配推算正常應力水準下的可靠度

d. 逐步加壓測試(Step-stress test)：於試驗開始時，對受測品施以正常操作時的環境應力，在經過一段時間後即將應力增加，如此逐次加壓直到受測品失效爲止的測試。故此種測試有時間及應力兩個參數。

e. 對早夭、隨機、老化等三種失效均適用。

(4) 對於已可掌握失效機制且可用數學式描述的情形最爲適用。

(5) Arrhenius 方程式：係用來描述與溫度相關的反應如金屬生鏽、潤滑劑失效等。

 a. 流程率(Rate of process) = $Ae^{-\frac{Q}{RT}}$

  其中 A：常數

   Q：活化能(Activation energy)

   R：氣體常數

   T：溫度(K)

 b. 設正常操作失效率爲 λ，加速測試失效率爲 λ_a，則

$$\frac{\lambda}{\lambda_a} = \frac{Ae^{-\frac{Q}{RT}}}{Ae^{-\frac{Q}{RT_a}}} = \exp\left[\frac{Q}{R}\left(\frac{1}{T_a} - \frac{1}{T}\right)\right] \quad\text{................................. (6.15)}$$

例 6.5 對某 CMOS IC 進行加速壽命測試，測試參數如下：

樣本量	300
測試時數	5000 小時
測試溫度	168℃
失效個數	5

設失效爲隨機失效，且符合 Arrhenius 方程式。請問：該受測品在 50℃時之失效率？

解 加速壽命測試之 MTTF $= \dfrac{300 \times 5000}{5} = 3 \times 10^5$ (小時)，

故加速測試失效率 $\lambda_a = \dfrac{1}{3 \times 10^5} = 3.33 \times 10^{-6}$ (次/小時)。

由 MIL-HDBK-217 可查得 CMOS 的 Q/R = 7532。另 168℃ = 441K，50℃ = 323K。由(6.15)式

$$\frac{\lambda}{\lambda_a} = \exp\left[\frac{Q}{R}\left(\frac{1}{T_a} - \frac{1}{T}\right)\right] = \exp\left[7532\left(\frac{1}{441} - \frac{1}{323}\right)\right] = 1.9508 \times 10^{-3}$$

$$\Rightarrow \quad \lambda = 3.33 \times 10^{-6} \times \exp\left[7532\left(\frac{1}{441} - \frac{1}{323}\right)\right] = 0.0065 \times 10^{-6} \quad \text{(次/小時)}$$

■ 6.6　參數法(Parametric method)

1. 首先須確定經測試所得之數據符合何種分配？若以圖示法(其他有統計計算法或以套裝軟體適配等方法)則係將未分組數據點繪於某種分配的機率紙上，若符合該分配，則數據點將呈現一直線。

2. 指數分配

   (1) 因為指數分配(或定值失效率模式)係唯一僅有一個參數待估計的分配，故通常最先被嘗試。

   (2) 若失效率為定值則點繪圖形呈一直線，若為隨時間遞增則呈向上彎曲 (Concave up)，若為隨時間遞減則呈向下彎曲(Concave down)。

   (3) 對數分配機率紙之 X 座標為失效時間，Y 座標為累計分配值 $F(t)$。

   (4) 設 n 為受測品個數，則 $\hat{F}(t_i) = \dfrac{i}{n+1}$。

   (5) ∵ $R(t) = e^{-\lambda t}$，設 $\lambda t = 1$，則 $R(t) = e^{-1} = 0.368 = 1 - F(t)$，

   ∴ $F(t) = 1 - 0.368 = 0.632$。故當 $F(t)=0.632$ 時，$\lambda t = 1$ $\Rightarrow$ $\lambda = \dfrac{1}{t}$

例 6.6 某受測品 8 個，其失效時間分別為 80，134，148，186，238，450，581，890。將數據描繪於指數分配紙上並估計其失效率。

解 由失效數據計算 $F(t)$ 所得如下表。

i	t_i(小時)	$\hat{F}(t_i) = \dfrac{i}{n+1}$
1	80	0.111
2	134	0.222
3	148	0.333
4	186	0.444
5	238	0.555
6	450	0.666
7	581	0.777
8	890	0.888

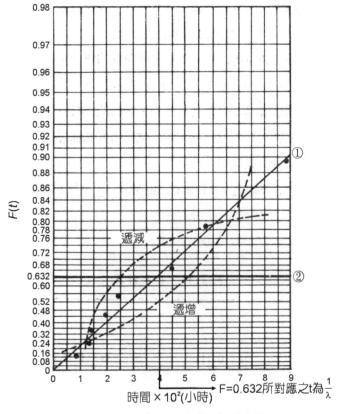

圖 6-10　指數分配的圖示參數估計

由圖 6-10 中 $F(t) = 0.632$ 所對應之 t ≈ 400 小時，

故估計 $\lambda = 1/400 = 0.0025$(次/小時)

3. 常態分配

$$F(t) = \Phi\left(\frac{t-\mu}{\sigma}\right) \quad\Rightarrow\quad F(\mu) = \Phi(0) = 0.5$$

$$\Rightarrow \quad F(\mu + \sigma) = \Phi(1) = 0.8413$$

例 6.7 某受測品 8 個，其失效時間分別為 13.7，18.1，19.3，20.6，22.3，23.2，26.5，31.0。將數據描繪於常態分配紙上並(1)由圖中找出 μ 及 σ 的估計值，(2)由數據計算 μ 及 σ 的估計值。

解 由失效數據計算 $F(t)$ 所得如下表。

i	t_i(小時)	$\hat{F}(t_i) = \dfrac{i}{n+1}$
1	13.7	0.111
2	18.1	0.222
3	19.3	0.333
4	20.6	0.444
5	22.3	0.555
6	23.2	0.666
7	26.5	0.777
8	31.0	0.888

將數據點繪於常態分配機率紙如圖 6-11。

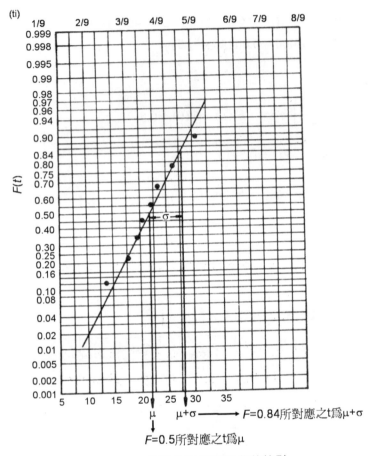

圖 6-11　常態分配的圖示參數估計

(1)　由圖 6-11 可得 $\hat{\mu} = 22$ 小時，$\hat{\sigma} = 28 - 22 = 6.0$ 小時

(2)　a.　由平均數的不偏估計量 $\bar{x} = \dfrac{1}{n}\sum_{i=1}^{n} x_i$，計算 $\hat{\mu} = \dfrac{174.7}{8} = 21.8$（小時）

　　　b.　由變異數的不偏估計量 $S_{\bar{x}}^{2} = \dfrac{1}{n-1}\sum_{i=1}^{n}\left(x_i - \bar{x}\right)^2$，計算 $\hat{\sigma}$

$$\hat{\sigma}^2 = \frac{n}{n-1}\left(\frac{1}{n}\sum t_i^2 - \hat{\mu}^2\right) = \frac{8}{7}\left[501.4 - (21.8)^2\right] = 29.9 \text{（小時）}$$

$$\Rightarrow \quad \hat{\sigma} = 5.5 \text{ 小時}$$

4. 對數常態分配

(1) 對數常態分配機率紙係將常態分配機率紙之橫軸改為對數值。

(2) 由常態分配 $F(t) = \Phi\left(\dfrac{t-\mu}{\sigma}\right)$ $\Rightarrow$ 對數常態分 $F(t) = \Phi\left(\dfrac{\ln t - \ln \mu}{\ln \sigma}\right)$

$\Rightarrow$ $\Phi^{-1}[F(t)] = \dfrac{1}{\ln \sigma} \times \ln t - \dfrac{\ln \mu}{\ln \sigma}$ 此為一條直線方程式，

其中 X 軸為 $\Phi^{-1}[F(t)]$，Y 軸為 $\ln t$，斜率為 $\dfrac{1}{\ln \sigma}$。

(3) $F(\mu) = \Phi\left(\dfrac{\ln \mu - \ln \mu}{\ln \sigma}\right) = \Phi(0) = 0.5$，故 $F(t) = 0.5$ 處所對應之 t 即為 μ。

(4) 量取圖形之斜率即為 $\dfrac{1}{\ln \sigma}$，再經計算則可得 σ。

例 **6.8** 某受測品 20 個，一半在 190°F 另一半在 240°F 的環境下進行壽命試驗。其失效時間(單位：分鐘)記錄如下：

1. 190°F：7228，7228，7228，8448，9167，9167，9167，9167，10511，10511，

2. 240°F：1175，1175，1521，1567，1617，1665，1665，1713，1761，1953。因該記錄係定時取樣的結果，故有些數據是相同的。將數據描繪於對數常態分配紙上並找出 μ 及 σ 的估計值。

解 以 $\hat{F}(t_i) = \dfrac{i}{n+1} = \dfrac{i}{11}$ 求取 Y 軸之值，以記錄之失效時間為 X 軸之值，點繪於對數常態分配紙上，結果如圖 6-12 所示。

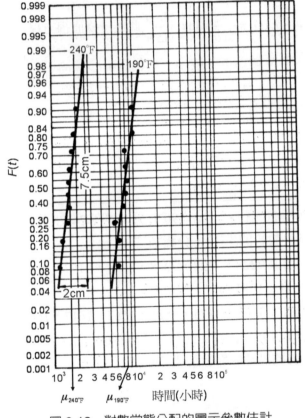

圖 6-12 對數常態分配的圖示參數估計

1. 由圖 6-12 中 F(t) = 0.5 所對應之值可直接讀取：$\hat{\mu}_{240°F} = 1.6 \times 10^3$ 分鐘，$\hat{\mu}_{190°F} = 8 \times 10^3$ 分鐘。

2. 由圖 6-12 中量得兩條直線之斜率均為 $\dfrac{7.5}{2} = \dfrac{1}{\ln\sigma}$ $\Rightarrow$ 估計 $\sigma = 1.31$ 分鐘

5. 韋氏分配

(1) 二參數型

　　a.　$R(t) = \exp\left[-\int_0^t \lambda(t)\,dt \right] = \exp\left[-\left(\dfrac{t}{\alpha}\right)^\beta \right]$

$\Rightarrow \quad \left(\dfrac{t}{\alpha}\right)^\beta = \ln\left(\dfrac{1}{R}\right)$，$\Rightarrow \quad \beta\ln\left(\dfrac{t}{\alpha}\right) = \ln\left[\ln\left(\dfrac{1}{R}\right)\right]$，

$\Rightarrow \quad \beta\ln t - \beta\ln\alpha = \ln\left[\ln\left(\dfrac{1}{R}\right)\right]$

再將 (1-F) 取代 R 後寫成直線方程式，得

$\ln\left[\ln\left(\dfrac{1}{1-F}\right)\right] = \beta\ln t - \beta\ln\alpha$，斜率為 β。

　　b.　若 $t = \alpha$，則 $\ln\left[\ln\left(\dfrac{1}{1-F}\right)\right] = \beta\ln\alpha - \beta\ln\alpha = 0$，

$\Rightarrow \quad \ln\left(\dfrac{1}{1-F}\right) = 1$，$\Rightarrow \quad \dfrac{1}{1-F} = e^1$，$\Rightarrow \quad F = \dfrac{e-1}{e} = 0.632$

例 6.9　某受測品 5 個，其失效時間分別為 67，120，130，220，290。將數據描繪於韋氏分配紙上並找出 α 及 β 的估計值。

解　由失效數據計算 $F(t)$ 所得如下表。

i	t_i(小時)	$\hat{F}(t_i) = \dfrac{i}{n+1}$
1	67	0.166
2	120	0.333
3	130	0.500
4	220	0.666
5	290	0.833

將數據點繪於韋氏分配機率紙如圖 6-13。

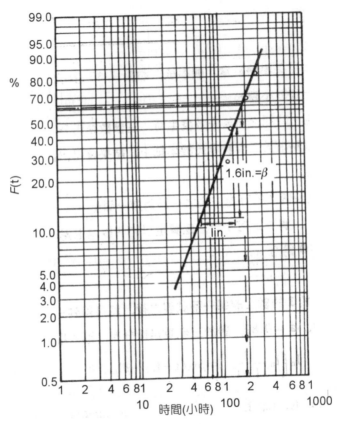

圖 6-13　韋式分配的圖示參數估計

由圖 6-13 中量得斜率，估計 $\beta = 1.6$，$\alpha = 190$ 小時。

(2) 三參數型

　　a.　因三 e 參數型係描述產品有一最低壽命 t_0 $(t_0 \neq 0)$，故公式應修正

　　　　為 $\ln\left[\ln\left(\dfrac{1}{1-F}\right)\right] = \beta \ln(t - t_0) - \beta \ln \alpha$，

　　　　i.e. 將二參數型韋氏分配的橫軸由 t 改為 $(t - t_0)$，斜率仍為 β。

b. 若直接將有非零最低壽命之數據點繪至韋氏分配機率紙，則會發生向下彎曲的情形，如例題 6.10 所示。

若令 t_0 等於第一個失效時間 t_1，並將韋氏分配的橫軸由 t 改為$(t - t_0)$可改善向下彎曲的非線性情形。但若令 $t_0 = at_1$，$0 < a < 1$，並以試誤法(Trail and error)求取最佳 a 值，則可得到一條近似直線。

例 6.10 某受測品 8 個，其失效時間分別為 220，251，285，346，361，585，646，1040。

1. 直接將數據描繪於韋氏分配紙上，
2. 以三參數型處理，並以試誤法求取最佳 a 值。

解

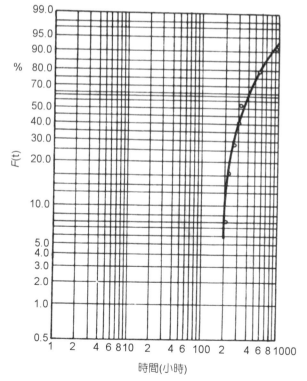

圖 6-14　韋式點繪指出非零最低壽命

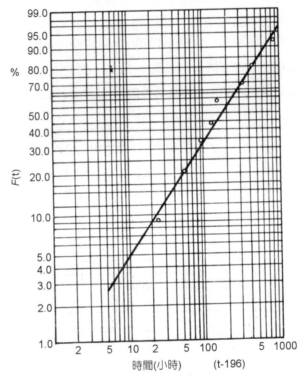

圖 6-15　韋式點繪調整非零最低壽命

經試誤後，取 $a = 0.89$ ，$t_0 = at_1 = 0.89 \times 220 = 196$ (小時)

■ 6.7 定值失效率估計

1. 定值失效率模式(指數分配)，因係描述浴缸曲線中有用壽命的部分(隨機失效)，故是可靠度工程中應用最廣的模式。

2. 判斷是否為定值失效率模式的方法至少有：

   (1) 引用已經證實為指數分配的類似產品，

   (2) 利用統計適配法，

(3) 利用指數分配機率紙的作圖法，

(4) 由失效模式判定其為隨機失效而非早夭失效亦非老化失效。

3. 自右截略

(1) 若測試品數量很多時，要完成所有受測品均失效的測試通常須很久的時間，而且最後幾個失效的受測品之失效模式中均含有老化的因素。所以可採取截略手段使老化因素自測試數據中剔除。

(2) 截略可分成 Type I(定時型)及 Type II(定數型)，而每型又可分成置換與不置換兩種類型。置換法(Replacement method)係指於測試執行中，若有受測品失效則立即更換或修復，保持受測品數目不變的方法。不置換法(Non-replacement method)則是將失效品自測試程序中取出，以致受測品數目逐漸減少的測試法。

(3) 定時型的好處是測試時間係於事前決定，故便於實驗的規劃；而定數型測試無法預知第 n 個失效發生的時間。但定數型卻可提供決定指數分配的 MTTF 量測值精度所需的失效個數，其並非測試時間的函數。選擇置換或不置換測試時須考慮成本因素，包括受測品成本與置換工作之人員費用等。

4. MTTF 估計值

設受測品總數為 n，其中 k 個失效品之失效時間分別為 t_1，t_2，……，t_k。

(1) 計時型：設約定的測試時間為 t_*。

    a.　不置換

       測試總時間為 $T = \sum_{i=1}^{k} t_i + (n-k) \times t_*$

例 6.11　某受測品 100 個一起測試 10 小時，測試期間若有失效則將失效品自測試程序中取出。現記錄得測試期間內共有 10 個受測品失效，其失效時間分別為 1，3，6，2，3，6，8，9，2，1 小時。請估計 MTTF 之值。

解 $T = \sum\limits_{i=1}^{k} t_i + (n-k) \times t_* = \sum\limits_{i=1}^{10} t_i + (100-10) \times 10 = 41 + 900 = 941$ (小時)

$\text{MTTF} = \hat{\mu} = \dfrac{T}{k} = \dfrac{941}{10} = 94.1$ (小時)

---

    b.   置換

       測試總時間為 $T = n \times t_*$

例 6.12 某受測品 24 個一起測試 5000 小時,測試期間若有失效則立即將失效品更新。現記錄得測試期間內共有 14 次失效發生,請估計 MTTF 之值。

解 $T = n \times t_* = 24 \times 5000 = 120{,}000$ (小時)

$\text{MTTF} = \hat{\mu} = \dfrac{T}{k} = \dfrac{120{,}000}{14} = 8571$ (小時)

---

(2)   計數型:約定於第 k 個失效發生即終止測試。

    a.   不置換

       測試總時間為 $T = \sum\limits_{i=1}^{k} t_i + (n-k) \times t_k$

例 6.13 某受測品 100 個一起進行測試,測試期間若有失效則將失效品自測試程序中取出,且決定第 11 個失效發生時即終止測試。現記錄得其失效時間分別為 1,1,2,2,3,3,6,6,8,9,15 小時。請估計 MTTF 之值。

解 $T = \displaystyle\sum_{i=1}^{11} t_i + (100 - 11) \times 15 = (41 + 15) + (100 - 11) \times 15 = 1391$ （小時）

$\text{MTTF} = \hat{\mu} = \dfrac{T}{k} = \dfrac{1391}{11} = 126.45$ （小時）

---

b. 置換

測試總時間為 $T = n \times t_k$

例 6.14 某受測品 6 個一起進行測試，測試期間若有失效則立即將失效品更新。現記錄得測試進行至 840 小時時就有 8 次失效發生，因測試者發覺設計有誤，故決定立即停止測試，依上述數據請估計該品之 MTTF 值。

解 $T = n \times t_k = 6 \times 840 = 5040$ （小時）

$\text{MTTF} = \hat{\mu} = \dfrac{T}{k} = \dfrac{5040}{8} = 630$ （小時）

---

5. MTTF 信任區間的查圖法

(1) Type II 定數法

a. 設 $U_{\frac{\alpha}{2},k}$ 及 $L_{\frac{\alpha}{2},k}$ 分別為平均數 μ 之 $100(1 - \alpha)\%$ 信任區間的上、下限，亦即：「以在第 k 個受測品失效時停止測試所得之數據，可估計 μ 的真值落於 $U_{\frac{\alpha}{2},k}$ 與 $L_{\frac{\alpha}{2},k}$ 之間的機率為 $(1 - \alpha)$」或

$\text{P}(L_{\frac{\alpha}{2},k} \leq \mu \leq U_{\frac{\alpha}{2},k}) = (1 - \alpha)$。而且 $\dfrac{L_{\alpha/2,k}}{\hat{\mu}}$ 與 $\dfrac{U_{\alpha/2,k}}{\hat{\mu}}$ 均與操作時間 T 無關，是顯著水準 α 及失效個數 k 的函數。若平均數以 $\hat{\mu} = \dfrac{T}{k}$ 估得，則 MTTF 信任區間的上下限 $U_{\frac{\alpha}{2},k}$ 與 $L_{\frac{\alpha}{2},k}$ 可由圖 6-16 查得。

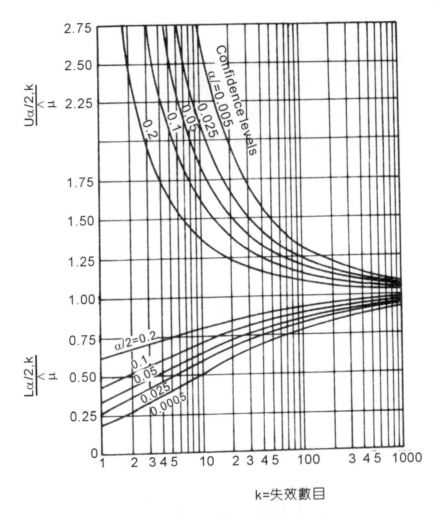

圖 6-16　MTTF 之信任界限

b. 若由數據僅求信任下限(單尾) $L_{\alpha,k}$，
則由圖 6-16 查 $\dfrac{L_{\alpha,k}}{\hat{\mu}}$ 之值後計算可得。

例 6.15 某受測品 20 個一起進行 30 小時不置換測試，於測試期間發生失效 9 次，且記錄得其失效時間分別為 14.4，5.1，27.7，29.1，23.6，20.0，10.5，13.5，27.4 小時。

1. 請估計 MTTF 之值，
2. 若此測試決定等第 10 個失效發生才終止，而第 10 個失效發生的時間為 41.2 小時，請估計 MTTF 之值，
3. 求(2)題中 MTTF 的 90%信任區間。

解

1. $T = \sum_{i=1}^{9} t_i + (20-9) \times 30 = 171.3 + 11 \times 30 = 501.3$ (小時)

2. $T = \sum_{i=1}^{10} t_i + (20-10) \times 41.2 = (171.3 + 41.2) + 01 \times 41.2 = 624.5$ (小時)

   $\hat{\mu} = \dfrac{T}{k} = \dfrac{624.5}{10} = 62.4$ (小時)

3. $\alpha = 0.1$，$\alpha/2 = 0.05$，$k = 10$，

   查圖 6-16 可得 $\dfrac{L_{0.05,10}}{\hat{\mu}} \approx 0.65$，$\dfrac{U_{0.05,10}}{\hat{\mu}} \approx 1.82$

   $\Rightarrow$　$U_{0.05,10} \approx 1.82 \times 62.4 = 114$ (小時)，$L_{0.05,10} \approx 0.65 \times 62.4 = 41$ (小時)

例 6.16 要求某產品之 90%MTBF 至少須為 100 小時。現取該產品一個進行壽命測試，其首次失效時間發生於第 210 小時。根據上述測試數據，該產品是否合乎要求？

解

解 $\alpha = 0.1$，$k = 1$，

$\hat{\mu} = \dfrac{T}{k} = \dfrac{210}{1} = 210$ (小時)

查圖 6-16 可得 $\dfrac{L_{\alpha,k}}{\hat{\mu}} = \dfrac{L_{0.1,1}}{\hat{\mu}} \approx 0.44$　$\Rightarrow$　$L_{0.1,1} \approx 0.44 \times 210 = 92.4$ (小時)，

故不合乎要求。

(2) Type I 定時法

設 $U^*_{\frac{\alpha}{2},k}$ 及 $L^*_{\frac{\alpha}{2},k}$ 分別為 Type I 定時法所求得平均數 μ^* 之 $100(1-\alpha)\%$信

任區間的上、下限，則由卡方分配的特性可知：

a. 定時法所求得之平均數與定數法所求得之平均數相等，

i.e. $\mu^* = \mu$。

b. 定時法所求得之信任上限與定數法所求得之信任上限相等，

i.e. $U^*_{\frac{\alpha}{2},k} = U_{\frac{\alpha}{2},k}$。

c. 定時法所求得之信任下限須修正為：$\dfrac{L^*_{\alpha/2,k}}{\hat{\mu}} = \left(\dfrac{k}{k+1} \times \dfrac{L_{\alpha/2,k}}{\hat{\mu}} \right)$。

d. 若由數據僅求信任下限(單尾) $L^*_{\alpha,k}$，則由圖 6-16 查 $\dfrac{L_{\alpha,k}}{\hat{\mu}}$ 之值後仍

須按上式修正為：$\dfrac{L^*_{\alpha,k}}{\hat{\mu}} = \left(\dfrac{k}{k+1} \times \dfrac{L_{\alpha,k}}{\hat{\mu}} \right)$，再經計算可得。

例 6.17 某產品一年內有 7 次失效記錄，求該產品 MTBF 的 95%信任區間。

解 解 $\alpha = 0.05$，$\alpha/2 = 0.025$，$k = 7$，

$\hat{\mu} = \dfrac{T}{k} = \dfrac{12}{7} = 1.71$(月)

$\dfrac{L^*_{\alpha/2,k}}{\hat{\mu}} = \left(\dfrac{k}{k+1} \times \dfrac{L_{\alpha/2,k}}{\hat{\mu}} \right)$，由圖 6-16 查得 $\dfrac{L_{\alpha/2,k}}{\hat{\mu}} = \dfrac{L_{0.05/2,7}}{\hat{\mu}} = 0.57$

$\Rightarrow \dfrac{L^*_{0.05/2,7}}{\hat{\mu}} = \left(\dfrac{7}{7+1} \times \dfrac{L_{0.05/2,7}}{\hat{\mu}} \right) = \dfrac{7}{8} \times 0.57 = 0.50$

$\Rightarrow L^*_{0.025,7} \approx 0.50 \times 1.71 = 0.86$ (月)

由圖 6-16 查得 $\dfrac{U_{\alpha/2,k}}{\hat{\mu}} = \dfrac{U_{0.05/2,7}}{\hat{\mu}} = 2.5$

$\Rightarrow \quad U_{0.025,7}^{*} = U_{0.025,7} = 2.5 \times 1.71 = 4.27 \,(月)$

例 6.18 要求某產品之 90%MTBF 至少須為 100 小時。現取該產品一個進行壽命測試，並計劃測試時間為 600 小時，期間若有失效發生則立即更換新品後繼續測試。

1. 若測試期間有 3 次失效，該產品是否合乎要求？
2. 若測試期間僅有 2 次失效，該產品是否合乎要求？

解 1. $\alpha = 0.1$，$k = 3$，

$\hat{\mu} = \dfrac{T}{k} = \dfrac{600}{3} = 200\,(小時)$，查圖 6-16 可得 $\dfrac{L_{\alpha,k}}{\hat{\mu}} = \dfrac{L_{0.1,3}}{\hat{\mu}} \approx 0.57$

$\Rightarrow \quad \dfrac{L_{0.1,3}^{*}}{\hat{\mu}} = \left(\dfrac{3}{3+1} \times \dfrac{L_{0.1,3}}{\hat{\mu}} \right) \approx \left(\dfrac{3}{4} \times 0.57 \right)$

$\Rightarrow \quad L_{0.1,3}^{*} \approx \dfrac{3}{4} \times 0.57 \times 200 = 85.5\,(小時)$，故不合乎要求。

2. $\alpha = 0.1$，$k = 2$，

$\hat{\mu} = \dfrac{T}{k} = \dfrac{600}{2} = 300\,(小時)$，查圖 6-16 可得 $\dfrac{L_{\alpha,k}}{\hat{\mu}} = \dfrac{L_{0.1,2}}{\hat{\mu}} \approx 0.53$

$\Rightarrow \quad \dfrac{L_{0.1,2}^{*}}{\hat{\mu}} = \left(\dfrac{2}{2+1} \times \dfrac{L_{0.1,2}}{\hat{\mu}} \right) = \left(\dfrac{2}{3} \times 0.53 \right)$

$\Rightarrow \quad L_{0.1,2}^{*} \approx \dfrac{2}{3} \times 0.53 \times 300 = 106\,(小時)$，故合乎要求。

■ 6.8　可靠度成長測試

1. 產品於雛型製作完成要量產之前，通常會經過一測試階段，目的在於經由測試以找出設計或製造時沒有注意到的缺陷，根據測試結果修正原來的設計，以得到較佳的產品品質並提高該產品的可靠度。這樣的測試很可能不

只一次，而是測試－修正－測試－修正……不斷反覆地進行以達到最佳結果。在這反覆的過程中，產品的缺陷不斷被發現並自設計中剔除，故產品的 MTBF 不斷地加長，可靠度亦不斷地提昇，此即所謂「可靠度成長」。

2. 可靠度成長測試又稱「雛型測試(Prototype test)」，其與「燒入測試(Burn-in test)」不同處在於：雛型測試是對產品不斷地修正，該產品的失效率因設計改良而遞減；而燒入測試是於測試過程中將失效品移除，測試仍繼續進行，但並未對設計進行改善。

3. 杜安法(Duane's method)

    (1) 係一種利用雛型測試數據來估計產品最終可靠度參數的方法，由杜安 (J. J. Duane)提出。

    (2) 設 T 為雛型測試時所有受測品累積之總操作時間，$n(T)$ 為 T 時間內之總失效次數，則 $n(T)/T$ 即為平均失效率。杜安觀察到將機電設備 (Electromechanical equipment)的 $n(T)/T$ 對 T 描繪於 log-log 座標上時，$\ln[n(T)/T]$ 對 $\ln(T)$ 均呈一直線，如圖 6-17 所示。此類圖形稱為杜安圖(Duane Plot)。由圖中可見失效率隨時間遞減，亦即 MTBF 隨時間遞增的現象(可靠度成長)。

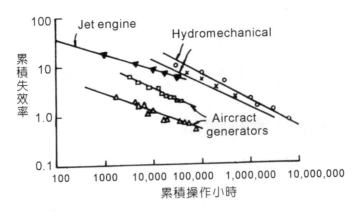

圖 6-17　杜安圖

(3) 設 $\lambda_i(t)$ 為瞬時失效率，亦即單位操作時間內的失效次數，故 $\lambda_i(t) \equiv \dfrac{d}{dt} n(t)$。將 $\lambda(t)$ 由 0 積分至 T 即為時間[0, T]的累計失效率 $\lambda_c(T)$，i.e. $\displaystyle\int_0^T \lambda_i(t)dt = n(T) = \lambda_c(T)$。

(4) 設 $\mu_i(T)$、$\mu_c(T)$ 分別為瞬時 MTBF 及平均累計 MTBF，則

$$\mu_i(T) \equiv \frac{1}{\lambda_i(T)} = \frac{1}{\dfrac{d}{dt} n(T)} \quad \dotfill (6.16)$$

$$\mu_c(T) \equiv \frac{1}{\dfrac{1}{T}\lambda_c(T)} = \frac{1}{\dfrac{1}{T} n(T)} = \frac{T}{n(T)} \quad \dotfill (6.17)$$

(5) 因為 $\ln[n(T)/T]$ 對 $\ln(T)$ 呈一直線，故可將兩者寫成直線方程式：$y = -\alpha x + b$，其中 $y = \ln[n(T)/T]$，$x = \ln(T)$，$\alpha = $ 直線斜率，i.e. $\ln[n(T)/T] = -\alpha \ln T + b$，

$$\Rightarrow \quad n(T)/T = e^{-\alpha \ln T + b} = e^b \times e^{-\alpha \ln T} = K e^{\ln T^{-\alpha}} = K T^{-\alpha}$$

$$\Rightarrow \quad n(T) = K T^{1-\alpha} \quad \dotfill (6.18)$$

(6) 由(6.16)及(6.18)可得　$\lambda_i(T) = \dfrac{d}{dt} n(T) = \dfrac{d}{dt} K T^{1-\alpha} = (1-\alpha)K T^{-\alpha}$

$$\Rightarrow \quad \mu_i(T) = \frac{1}{\dfrac{d}{dt} n(T)} = \frac{1}{(1-\alpha)K} T^{\alpha} \quad \dotfill (6.19)$$

寫成直線方程式

$$\ln[\mu_i(T)] = \frac{1}{(1-\alpha)K} \ln[T^{\alpha}] = \frac{\alpha}{(1-\alpha)K} \ln[T]$$

(7) 由(6.17)及(6.18)可得

$$\mu_c(T) = \frac{T}{n(T)} = \frac{T}{KT^{1-\alpha}} = \frac{1}{K}T^{\alpha} \text{..} (6.20)$$

寫成直線方程式

$$\ln\left[\mu_c(T)\right] = \frac{1}{K}\ln\left[T^{\alpha}\right] = \frac{\alpha}{K}\ln\left[T\right]$$

(8) 由(6.19)及(6.20)可得

$$\mu_i(T) = \frac{1}{(1-\alpha)}\mu_c(T) \text{..} (6.21)$$

(9) $\mu_i(T)$、$\mu_c(T)$ 的杜安圖如圖 6-18 所示。

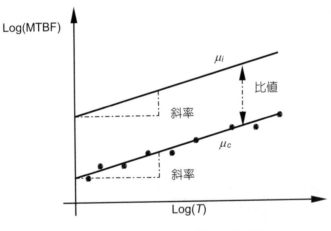

圖 6-18　$\mu_i(T)$、$\mu_c(T)$ 的杜安圖

4. 討論

(1) 可由圖 6-18 以外差法估計可靠度的成長。

(2) 若失效係源自無法移去之設計缺陷,則可靠度成長的效果將不顯著。

(3) 因為隨著修正設計次數的增加，邊際可靠度是遞減的，故愈往後為提高可靠度而付出費用的投資報酬率是否合算？特別是為減少隨機失效或老化磨耗失效的支出，值得作成本分析。

(4) 量產品的可靠度通常會低於雛型的可靠度，產品操作於現場的可靠度通常會低於操作於實驗室時的可靠度，計算可靠度的成長時，這些因素亦應納入考量。

■ 6.9　預防保養與失效預測

產品的可靠度與產品的失效息息相關，若產品持續不失效地在可操作(Functioning)狀態，則產品即持續在可靠的狀態。適當的維修保養(Maintenance)是防止或延緩產品失效的有效方法。維修保養的執行時間在失效發生之後者稱為修正保養(Corrective Maintenance, CM)，在失效發生之前者則稱為預防保養(Preventive Maintenance, PM)。對於失效發生並不會造成嚴重後果的一般產品，通常施以修正保養，例如教室照明燈管燒壞、汽車雨刷損壞等。然對於失效會造成嚴重後果者，或許是重大經濟損失例如晶圓廠的設備失效，或是嚴重生命傷亡例如飛機故障失事、核電廠爆炸等，則應施以預防保養以期在失效發生之前即將造成失效的因素消除，進而使失效不會發生。但是預防保養最困難的地方是「人類無法預知失效何時發生？」故失效預測(Failure prediction)一直是學術界與工程界努力的課題。現行的預防保養分成兩大類：

1. 時基式(Time-based)，

2. 狀態基式(Condition-based)。

時基式是依過去該產品的維修經驗，訂定一維修保養時間週期(或是使用次數、行駛里程等等)，週期一到即執行維修保養，電力公司的歲修即是一例。但是保養週期若訂得短些，雖可降低遭遇失效發生的機率，卻會增加成本的支出；反之，保養週期若訂得太長，雖可節省開銷但卻使遭遇失效的機率大增而失去了預防保養的意義。因為保養週期的訂定係根據過去的經驗和記錄統計而

得，所以會有兩個方向的缺失：

1. 產品實際壽命可能長於保養週期，但卻於維修保養時被更換，如此造成投資浪費。

2. 產品實際壽命可能短於保養週期，故仍有在定期維修保養前發生失效的風險。

狀態基式的預防保養是根據產品實際的狀態來決定執行維修保養計劃的方法，狀態基式可以免除上述時基式的缺點，所以應是較佳的維修保養策略。狀態基式預防保養的研討和實例請參考附錄 C、E。

❖ 習 題

1. 四個產品的 TTF 分別為：240，420，630，1080，單位為小時。求：

   (1) 將數據點會於對數常態分配機率紙上並觀察其結果，

   (2) 以此結果估計 MTTF。

2. 十個產品的燒入時間分別為：17.0，20.6，29.7，27.7，26.5，27.0，25.6，21.4，22.7，21.3，單位為分鐘。請以非參數法及參數法分別求該產品的可靠度和失效率。

3. 請繪圖並說明你對 Condition-based Preventive Maintenance 有何策略與想法？

4. 請設計一失效預測的方法與實驗。

附 錄

Appendix

附錄 A

A PETRI-NET APPROACH TO FINDING MINIMUM TIE-SETS FOR SYSTEM RELIABILITY EVALUATION

SUMMARY

A system's reliability must be evaluated after it is configured so as to judge whether the design is acceptable or not. A method for system reliability evaluation based on Petri nets is introduced in this study. This method starts from transforming the reliability block diagram of a system to be evaluated to a Petri net. The followed step is to establish a matrix in accordance with the Petri net, which gives all minimum tie-sets of the system. Finally, the reliability of the system is obtained from the union of the probabilities of the minimum tie-sets. This method offers a logical way to find out all minimum tie-sets of a system, from which the reliability of the system is resulted. Three examples are given to demonstrate this method and to compare with other three well-known methods for system reliability evaluation. Moreover, a power distribution system is employed to illustrate that how a complex-system reliability is evaluated by this method. As long as the schematic description of a system is given, the method can be applied to evaluate the reliability of the system. This method offers another option for system reliability evaluation.

∗ **KEY WORDS**： Hybrid Petri nets; Reliability evaluation; Tie-sets; Power distribution system

1. INTRODUCTION

A system (or a product) is a collection of components arranged according to a specific design in order to achieve desired functions with acceptable performance and reliability measures [1]. The type of components used, their qualities, and the design configuration in which they are arranged have a direct effect on the system performance and its reliability. There are various system configurations such as series, parallel, series-parallel or parallel-series, or complex networks. Once the system is configured, its reliability must be evaluated so as to judge whether the design is acceptable or not.

Among various system configurations, evaluating reliabilities for series or parallel type systems are relatively easier. However, special skills will be employed to evaluate the reliabilities of those complex systems that cannot be modeled (or are difficult to model) into series or parallel configurations. There have been many methods proposed for evaluating reliability of a complex network such as [1-4] decomposition method, tie-set and cut-set methods, Boolean truth table method, reduction method, factoring algorithm, or fuzzy set theory [5] etc. There is a common property for the above-mentioned methods, that is, the evaluating process will become quite labored and tedious [1] when the evaluated system is slightly large or complex. In the current study, a method for evaluating system reliability based on Petri nets is introduced. It is done by transforming the reliability block diagram of a system (which is easily constructed only if the schematic description of the system is given) to the correlated Petri net and then deriving a matrix, called a path matrix, from the Petri net according to the rules proposed in this study. Each column of the path matrix becomes a minimum tie-set of the system and the reliability of the evaluated system is given by the union of the probabilities of these minimum tie-sets. A similar method proposed by Liu and Chiou [6] is for failure analysis; however, the fault tree of the system has to be established beforehand. Another method-the association of stochastic Petri nets (SPNs) and Markov's chains constitute a powerful tool for analysis [7], which is for the situations that the random time-variable is participated. Nevertheless, the Markovian models obtained for complex systems are large and complicated to deal with. By contrast, the matrix method proposed in this study offers a logical way, no extra pre-processing efforts needed, to find out all minimum tie-sets of a complex system such that the reliability of the system can be evaluated logically. This method offers another option for system reliability evaluation.

2. PETRI NETS

A Petri net is a general-purpose mathematical tool for describing relations existing between conditions and events [8]. The basic symbols of Petri nets include [9]:

○ : Place, drawn as a circle, denotes event

—— : Immediate transition, drawn as a thin bar, denotes event transfer with no delay time

━━ : Timed transition, drawn as a thick bar, denotes event transfer with a period of delay time

↑ : Arc, drawn as an arrow, between places and transitions

● : Token, drawn as a dot, contained in places, denotes the data

Q : Inhibitor arc, drawn as a line with a circle end, between places and transitions

The basic structures of the logic relations for Petri nets are listed in Figure 1, where P, Q, and R are different events. There are two types of input places for transition in Figure 1; namely specified and conditional [10]. The former has a single output arc, whereas the latter has multiples. Tokens in a specified-type place have only one outgoing destination; that is, if the input place(s) holds a token then the transition fires and gives the output place(s) a token. However, tokens in a conditional-type place have more than one outgoing path, which may lead the system to different states. For the 'TRANSFER OR' Petri nets in Figure 1, whether Q or R takes over a token from P depends on conditions such as probability, extra action, or self-condition of the place. The symbol Q' in the 'INHIBITION' structure represents inverse Q.

There are three types of transitions that are classified based on time [8]. Transitions with no time delay due to transition are called immediate transitions, while those that need a certain constant period of time for transition are called timed transitions. The third type is called a stochastic transition and is used for modeling a process with random time. Owing to the variety of Petri nets, it is a powerful tool for modeling flexible manufacture systems, multilevel hierarchical systems, multiprocessor systems, etc [10]. Besides, Petri nets are capable not only of simulation, reliability analysis [11] and failure monitoring but also of dynamic behavior observation [12] and acts prediction [13].

Logic relation	TRANSFER	AND	OR	TRANSFER AND	TRANSFER OR	INHIBITION
Description	If P then Q	If P AND Q then R	If P OR Q then R	If P then Q AND R	If P then Q OR R	If P AND Q' then R
Boolean function	Q=P	R=P*Q	R=P+Q	Q=R=P	Q+R=P	R=P*Q'
Petri nets						

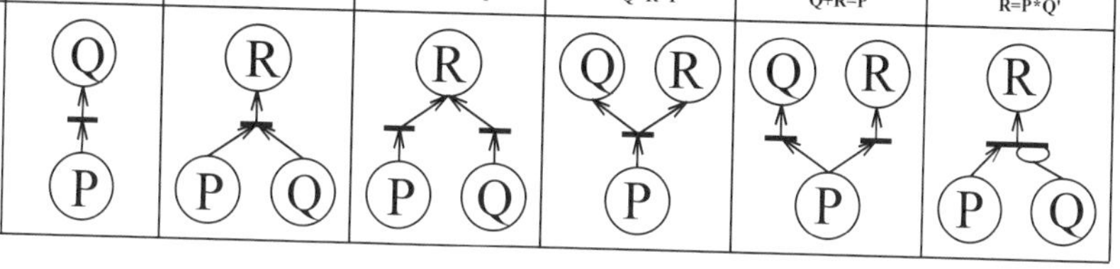

Figure 1. Basic structures of logic relations for Petri nets

3. MATRIX METHOD

It is a method for finding minimum tie-sets of a system. The reliability of a system is given by the union of the probabilities of the minimum tie-sets.

3.1 Transformation Between Reliability Block Diagram and Petri Net

The reliability block diagram of a system is a graphical representation of the components of the system and how they are connected. Figure 2 is an example of a reliability block diagram representing a mixed-parallel system [4]. The reliability block diagram is usually equivalent to the schematic description of a system. Therefore, the reliability block diagram of a system is inevitably given when the reliability evaluation problem of the system is initiated. If the components' failures in Figure 2 are independent, the reliability (R) of the system is expressed as

$$R=\{1-[1-P(A)P(B)][1-P(C)P(D)]\}\{1-[1-P(E)][1-P(F)]\}, \qquad (1)$$

where $P(X)$ is the probability that component X, $X = A, B, C, D, E, F$, is operational. If the components are identical and $P(X)=P$, then Eq. (1) can be rewritten as

$$R=[1-(1-P^2)^2][1-(1-P)^2]= 4P^3-2P^4-2P^5+P^6. \qquad (2)$$

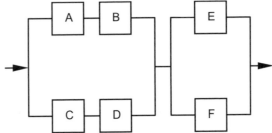

Figure 2.   Reliability block diagram for a mixed-parallel system

The first step in using matrix method to evaluate a system's reliability is to transform the reliability block diagram to a Petri net. The transformation rules are stated as follows:

1. Adding a dummy block to the input terminal and output terminal of the block diagram and give each of the blocks a symbol "I" or "O" to represent input and output, respectively.

2. A block and a path of the block diagram are transformed to a place and an arc of the Petri net, respectively.

3. Transforming it to a "TRANSFER OR" configuration as depicted in Figure 1 if any one block has multiple outgoing paths.

4. If a block has more than one input then the transformed place should be reproduced to the same number as the inputs so as to correspond to those inputs individually.

5. Starting from the I-block towards the O-block, construct the correlated Petri net step by step in a down-top fashion. This results in the correlated Petri net. Figure 3 is the correlated Petri net transformed from Figure 2.

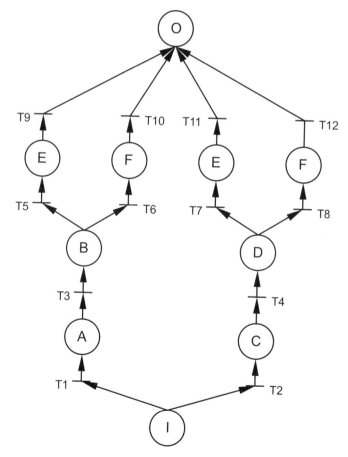

Figure 3. Petri net of Figure 2

3.2 Minimum Tie-Sets

A tie-set is a complete path through the block diagram of a system. In this study, the reliability block diagram is concerned. For a complex system, some of the tie-sets are contained within others. A minimum tie-set is the tie-set that contains no other tie-sets within it. The reliability of the system is given by the union of the probabilities of all minimum tie-sets [1]. Therefore, the reliability of a complex system can be determined if all minimum tie-sets of the system can be found.

By intuition, the minimum tie-sets of the mixed-parallel system depicted in Figure 2 are

$$T_1=ABE,$$
$$T_2=ABF,$$
$$T_3=CDE,$$
$$T_4=CDF. \tag{3}$$

Thus, the reliability of the system is

$$R=P(ABE \cup ABF \cup CDE \cup CDF). \tag{4}$$

3.3 Path Matrix

A matrix, namely path matrix, can be established in accordance with the correlated Petri net to find minimum tie-sets. The establishing method is described as follows.

1. Write down the names of the places by making a horizontal arrangement if the places are at the same level of the hierarchical structure of the Petri net.

2. If a place has more than one output arc, i.e. a TRANSFER OR configuration, the name of the place should be reproduced to the same number as the outputs then put down the reproduced names in the same row under the names of its output places.

3. All the place-names along the input arc route of the reproduced place should be reproduced to the same number as the reproduced places.

4. Following the above rules, starting from the places next lower than the O-place down to the places next higher than the I-place, the path matrix is thus formed.

Each column of the path matrix is a minimum tie-set and all minimum tie-sets of the system are revealed in the matrix. Figure 4 shows the path matrix of Figure 3. It gives all the minimum tie-sets from every column, i.e. ABE, ABF, CDE and CDF, which are the same as the result of Eq. (3). Consequently, the reliability of the system is

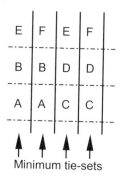

Figure 4.   Path matrix of Figure 3

$$R = P(ABE \cup ABF \cup CDE \cup CDF)$$
$$= P(ABE)+P(ABF)+P(CDE)+P(CDF)- [P(ABEF)$$
$$+P(ABCDE)+P(ABCDEF)+P(ABCDEF) +P(ABCDF)$$
$$+P(CDEF)]+[P(ABCDEF)+ P(ABCDEF)$$
$$+P(ABCDEF)+P(ABCDEF)]-P(ABCDEF). \qquad (5)$$

Assuming independence and equality of probabilities, Eq. (5) can be written as

$$R = 4P^3 -[P^4+ P^5+ P^6+ P^6+ P^5+ P^4]+[P^6+ P^6+ P^6+ P^6]- P^6$$
$$= 4P^3 -2P^4 -2P^5 +P^6. \qquad (6)$$

Obviously, Eq. (6) and Eq. (2) are identical.

4. EXAMPLES

The following examples are used to exemplify the matrix method and to compare with other three well-known methods for system reliability evaluation.

4.1 Comparing with Decomposition Method

Decomposition method begins by selecting a keystone component, u, which appears to link together the reliability structure of the system. The reliability may then be expressed in terms of the keystone component based on the theorem of total probability as [1]

$$R=P(\text{system good}/u)P(u) + P(\text{system good}/\bar{u})P(\bar{u}),$$

where P(system good/u) is the probability that the system is functioning given that u is functioning and P(system good/ū) is the probability that the system is functioning given that u is not functioning. As an example, to determine the reliability of the network shown in Figure 5 by decomposition method when B is selected as the keystone, the reliability is derived as [1]:

$$R = [P(D)+P(E)-P(D)P(E)]P(B)+[P(A)P(D)+P(C)P(E)-$$
$$P(A)P(D)P(C)P(E)][1-P(B)]. \tag{7}$$

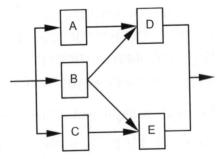

Figure 5. An illustrative network for decomposition method

If all components have equal probabilities P of functioning properly, then Eq. (7) becomes

$$R = [P+ P- P^2]\ P+ [P^2+ P^2- P^4]\ [1- P]$$
$$= 4P^2-3P^3-P^4+P^5.$$

(8)

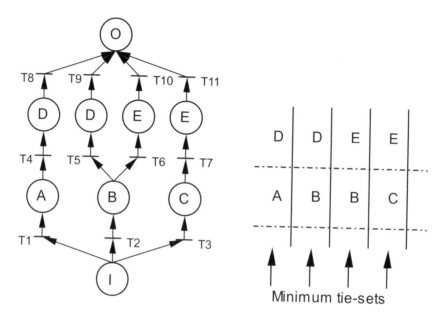

Figure 6.   Petri net of Figure 5   Figure 7.   Path matrix of Figure 6

To solve the same problem by the matrix method is demonstrated below. Figure 6 is the correlated Petri net transformed from Figure 5, and the corresponding path matrix is shown in Figure 7. All minimum tie-sets are found in Figure 7. Thus, the reliability of Figure 5 is expressed as:

$$R = P(AD\cup BD\cup BE\cup CE) = 4P^2-[P^3+ P^4+ P^4+ P^3+ P^4+ P^3]$$
$$+[P^4+ P^5+ P^5+ P^4]- P^5 = 4P^2-3P^3-P^4+P^5.$$

(9)

The results of Eqs. (8) and (9) are identical.

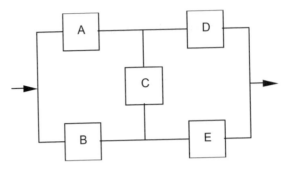

Figure 8.   An illustrative network for Boolean truth table method

4.2   Comparing with Boolean Truth Table Method

Boolean truth table method is based on the construction of a Boolean truth table for the system. A truth table lists every possible state of the system. A value of 1 or 0 is assigned to indicate that the component is functioning or not, respectively. The probability for every functioning state is computed and the reliability of the system is obtained by adding all functioning state probabilities [1]. Considering a network shown in Figure 8, the Boolean truth table for the network is constructed as shown in Table 1. Assume that the components are independent, are identical, and have the same probability P of operating properly, the reliability of the system is

$$R= P^5+ 5P^4(1-P)+8\ P^3(1-P)^2+2\ P^2(1-P)^3 = 2P^2+2P^3-5P^4+2P^5. \qquad (10)$$

Table 1. Boolean truth table for Figure 8

A	B	C	D	E	System	State Probability
1	1	1	1	1	1	P^5
1	1	1	1	0	1	$P^4(1-P)$
1	1	1	0	1	1	$P^4(1-P)$
1	1	1	0	0	0	
1	1	0	1	1	1	$P^4(1-P)$
1	1	0	1	0	1	$P^3(1-P)^2$
1	1	0	0	1	1	$P^3(1-P)^2$
1	1	0	0	0	0	
1	0	1	1	1	1	$P^4(1-P)$
1	0	1	1	0	1	$P^3(1-P)^2$
1	0	1	0	1	1	$P^3(1-P)^2$
1	0	1	0	0	0	
1	0	1	1	1	1	$P^3(1-P)^2$
1	0	0	1	0	1	$P^2(1-P)^3$
1	0	0	0	1	0	
1	0	0	0	0	0	
0	1	1	1	1	1	$P^4(1-P)$
0	1	1	1	0	1	$P^3(1-P)^2$
0	1	1	0	1	1	$P^3(1-P)^2$
0	1	1	0	0	0	
0	1	0	1	1	1	$P^3(1-P)^2$
0	1	0	1	0	0	
0	1	0	0	1	1	$P^2(1-P)^3$
0	1	0	0	0	0	
0	0	1	1	1	0	
0	0	1	1	0	0	
0	0	1	0	1	0	
0	0	1	0	0	0	
0	0	0	1	1	0	
0	0	0	1	0	0	
0	0	0	0	1	0	
0	0	0	0	0	0	

Figure 9 is the correlated Petri net transformed from Figure 8, and the corresponding path matrix is shown in Figure 10. By the matrix method, the reliability of the system depicted in Figure 8 is

$$R=P(AD\cup ACE\cup BCD\cup BE) = 2P^2+2P^3- [P^4+ P^4+ P^4+ P^5+ P^4+ P^4]+$$
$$[P^5+ P^5+ P^5+ P^5]- P^5 = 2P^2+2P^3-5P^4+2P^5. \tag{11}$$

Clearly, Eq. (10) and Eq. (11) are the same.

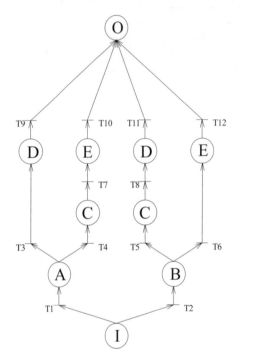

Figure 9. Petri net of Figure 8

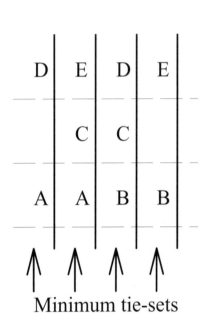

Minimum tie-sets

Figure 10. Path matrix of Figure 9

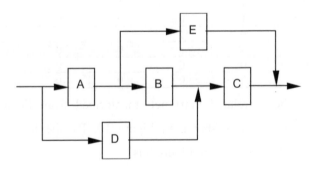

Figure 11. An illustrative network for tie-set method

4.3 Comparing with Tie-Set Method

Tie-set method is similar to the matrix method. However, the matrix method finds minimum tie-sets logically whereas tie-set method finds them by intuition. Considering the system shown in Figure 11, minimum tie-sets, AE, ABC, and DC, are found intuitively. By contrast, the matrix method finds minimum tie-sets by the path matrix shown in Figure 12 that is derived from the Petri net depicted in Figure 13. Hence, the reliability of the system is

$$R=P(AE \cup ABC \cup DC)=2P^2+P^3-3P^4+P^5. \tag{12}$$

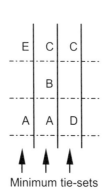

Figure 12.   Path matrix of Figure 13

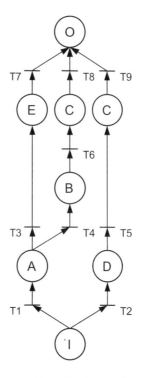

Figure 13.   Petri net of Figure 11

4.4 A Power distribution System

Figure 14 shows a simplified network of a power distribution system for a city [1]. The power is sent out from node 1 through transmission lines, A, B, C, D, E, F, G, and H, to distribution centers, node 2, 3, 4, 5, and 6. Power must reach node 6 since it provides electricity to critical services of the city. Thus, the reliability of the network is the probability that power sent from node 1 reaches node 6. Without a doubt, it is not easy to find out all minimum tie-sets of the system by intuition. Evaluating the reliability of the system by Boolean truth table method needs to deal with 256 states. By decomposition method, after selecting a keystone component, the decomposed portion is still quite complicated. In other words, it is not easy to determine the reliability of the system by using the above three methods.

On the other hand, the matrix method offers a logical way to evaluate the reliability. Figure 14 can be transformed to a Petri net as shown in Figure 15. Minimum tie-sets are obtained from the path matrix as shown in Figure 16. They are ADG, ADFH, ACEFG, ACEH, BCDG, BCDFH, BEFG, and BEH. Consequently, the reliability is given as

$$R = P(ADG \cup ADFH \cup ACEFG \cup ACEH \cup BCDG \cup BCDFH \cup BEFG \cup BEH)$$
$$= P\left[1 - (\overline{ADG} \cap \overline{ADFH} \cap \overline{ACEFG} \cap \overline{ACEH} \cap \right.$$
$$\left. \overline{BCDG} \cap \overline{BCDFH} \cap \overline{BEFG} \cap \overline{BEH})\right] \qquad (13)$$

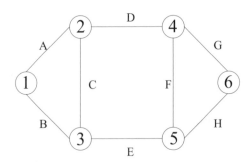

Figure 14. Simplified network of a power distribution system

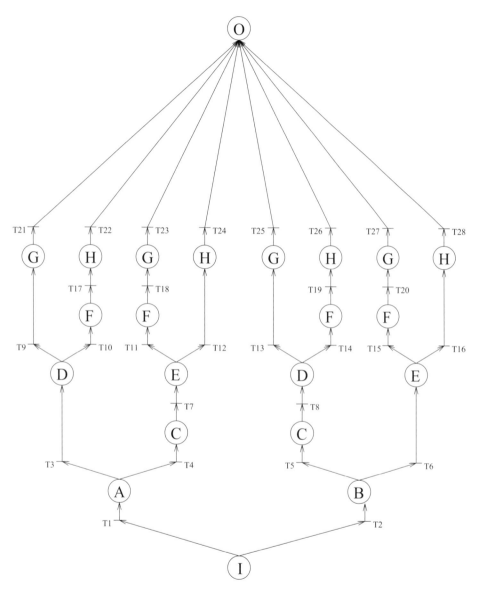

Figure 15.   Petri net of Figure 14

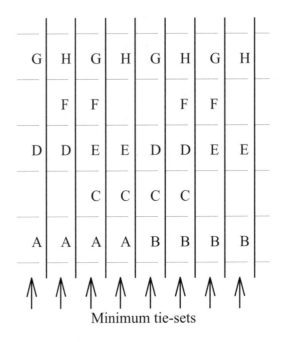

Minimum tie-sets

Figure 16.　Path matrix of Figure 15

5. DISCUSSIONS

1. Although there are many well-known methods for system reliability evaluation, they become labored and tedious when the system is large or complicated.

2. As long as the schematic description of a system is given, the matrix method can be applied to evaluate the reliability of the system. In case a system's reliability needs to be evaluated, of course, the schematic description for he system should be given to understand how the components of the system are connected. In other words, unlike the fault-tree method and the SPNs method, the matrix method needs no extra pre-processing efforts to perform the reliability evaluation.

3. The calculation for Eq. (13) can be done by a computer.

4. The matrix method offers a logical way to find out all minimum tie-sets of a system, and achieve the same result as the results achieved by the aforementioned well-known methods.

5. The Petri net transformed from a reliability block diagram offers not only a path matrix that gives all minimum tie-sets of the system but also the operation time needed for each minimum tie-set. The operation time is calculated by adding all transfer time of transitions of the minimum tie-set [12].

6.   CONCLUSIONS

A matrix method for system reliability evaluation based on Petri nets has been presented in this study. This method starts from transforming the reliability block diagram of a system to be evaluated to a Petri net. The followed step is to establish a path matrix in accordance with the Petri net, which gives all minimum tie-sets of the system. Finally, the union of the probabilities of the minimum tie-sets gives the reliability of the system. Three examples are given to demonstrate this method and to compare with other three well-known methods for system reliability evaluation. The results derived from the methods are the same. Moreover, a power distribution system is employed to illustrate that how a complex-system reliability is evaluated by this method. The matrix method offers not only a logical way to find out all minimum tie-sets of a complex system such that the reliability of the system can be evaluated logically, but also the operation time needed for each minimum tie-set. This method offers another option for system reliability evaluation.

REFERENCES

[1] E. A. Elsayed, *Reliability Engineering*, Addison Wesley Longman, Taipei, 1996.

[2] K.C. Kapur and L.R. Lamberson, *Reliability in Engineering Design*, Wiley, New York, 1977.

[3] B.S. Dhillon, *Mechanical Reliability: Theory, Models and Applications*, AIAA Education Series, Washington DC, 1988.

[4] S. S. Rao, *Reliability-Based Design*, McGraw-Hill, New York, 1992.

[5] A. R. Abdelaziz, A fuzzy-based power system reliability evaluation, *Electr. Power Syst. Res.*, 50 (1999) 1-5.

[6] T.S. Liu and S.B. Chiou, The application of Petri nets to failure analysis, *Reliability engineering and system safety*, 57 (1997) 129-142.

[7] Z. Simeu-Abazi, O. Daniel and B. Descotes-Genon, Analytical method to evaluate the dependability of manufacturing system, *Reliability engineering and system safety*, 55 (1997) 125-130.

[8] R. David and H. Alla, Petri nets for modeling of dynamic systems-A survey, *Automatica*, 30 (2) (1994) 175-202.

[9] W. G. Schneeweiss, *Petri net for reliability modeling*, Lilole-Verlag, Hagen, 1999.

[10] S. K. Yang and T. S. Liu, A Petri net approach to early failure detection and isolation for preventive maintenance, *Quality and Reliability Engineering International*, 14 (5) (1998) 319-330.

[11] K. Barkaoui, G. Florin, C. Fraize and B. Lemaire, Reliability analysis of non repairable systems using stochastic Petri nets, *Proc. 8th International Symposium on Fault-Tolerant Computing. Digest of Papers.* FTCS-18, 1988, pp. 90-95.

[12] S. K. Yang and T. S. Liu, Failure analysis for an airbag inflator by Petri nets, *Quality and Reliability Engineering International*, 13 (2) (1997) 139-151.

[13] Y. Manabe, M. Hattori, S. Tadokoro and T. Takamori, A model of human actions by a Petri net and prediction of human acts, *Transactions of The Japan Society of Mechanical Engineers*, 63 (1997) 287-294.

附錄 B

A PETRI-NET APPROACH
TO REMOTE DIAGNOSIS FOR
FAILURES OF CARDIAC PACEMAKERS

SUMMARY

This paper describes the application of Petri nets to remote diagnosis for failures of cardiac pacemakers. The operations, structures and basic control methods of the different types of cardiac pacemakers are first described. A combined synchronous pacemaker is modeled into a Petri net in this study. Twelve checkpoints are added into the modeled Petri net so as to construct a Petri net for failure diagnosis. A remote mode for failure diagnosis of implanted pacemakers is also designed by Petri net approach. A low-power transmitter transmits a checking-code with 12 digits from the implanted pacemaker to the outside of the patient's body manually or automatically. By observing the markings of the checking-code, the working status and the health condition of the pacemaker are clear at a glance. Applications of the Petri net method for failure diagnosis and control optimization are discussed.

∗**KEY WORDS**： cardiac pacemaker; Petri nets; failure diagnosis; remote control

1.   INTRODUCTION

A failure is defined as any change in the shape, size, or material properties of a structure, machine, or component that renders it unfit to carry out its specified function adequately [1]. For the purpose of reliability assurance, failures of a system need to be traced and analyzed especially for safety devices such as cardiac pacemakers and other therapeutic and prosthetic devices for human beings.

There have been many methods proposed for failure analysis [2], among which fault tree analysis (FTA) is well known. It is a graphical method that presents relationships between basic events and the top event by logic gates and a tree construction [3]. Compared with FTA, Petri net analysis is also a graphical approach that performs not only the static logic relations revealed in FTA, but also dynamic behaviour which greatly helps fault tracing and failure state analysis. Moreover, the system behaviour accounted for by Petri nets can improve the dialogue between analysts and designers of a system [4].

A cardiac pacemaker is an electric stimulator that produces periodic pulses that are conducted to electrodes located on the surface of the heart (the epicardium), within the heart muscle (the myocardium), or within the cavity of the heart or the lining of the heart (the endocardium). The stimulus thus conducted to the heart causes it to contract; this effect can be used prosthetically in disease states in which the heart is not stimulated at a proper rate on its own [5]. The principal pathologic conditions in which cardiac pacemakers are applied are known collectively as heart block [6].

2. PACEMAKERS

The package of an implanted pacemaker not only must be compatible and well tolerated by the body, but it must also provide the necessary protection to the circuit components in order to ensure their reliable operation. The body is a corrosive environment, so the package must be designed to operate well in this environment, while having minimum volume and mass. Early implanted pacemakers were encapsulated in a cast-epoxy resin, the external surface of which was often coated with a layer of silicone rubber. Silicon rubber is one of the best materials currently available in terms of biological compatibility with soft tissues. The lead wires and electrodes are, therefore, also frequently molded in silicone rubber to provide maximum biocompatibility for the critical locations in which they are found. Nowadays, cardiac pacemakers are packaged in hermetically sealed metal packages. Titanium and stainless steel are frequently used for the package. Special electron beam and laser welding techniques have been developed to seal these packages without damaging the electronic circuit or the power source. These metal packages take up less volume than the earlier polymer-based packages. [5]

The most frequently used power supply for implanted pacemakers is a battery made up of primary (non-rechargeable) cells. In the early development of the pacemaker, this battery consisted of high-quality zinc-mercury cells connected in series to serve as a 6V to 9V source. Although these cells were more reliable and had a greater energy density than the flashing batteries, they were the main cause of pacemaker failure and made it necessary to replace a pacemaker. Although special, high-reliability batteries are constructed for this application, customary practice in the late 1990s is to change the battery of a pacemaker every two to five years.

Cardiac pacemakers are either of the unipolar or bipolar type. In a unipolar one, a single electrode is in contact with the heart, and negative-going pulses are connected to it from the generator. A large indifferent electrode is located somewhere else in the body, usually mounted on the generator, to complete the circuit. In the bipolar system, two electrodes are placed on the heart, and the stimulus is applied across these electrodes. Both systems of electrodes require approximately the same stimulus for efficient cardiac pacing. Figure 1 shows the basic structures of the two types [5].

Different control algorithms lead to different configurations of pacemakers that are used for different heart-blocked patients. According to the control algorithms, pacemakers can be classified to asynchronous pacemakers, synchronous pacemakers, and rate-responsive pacemakers.

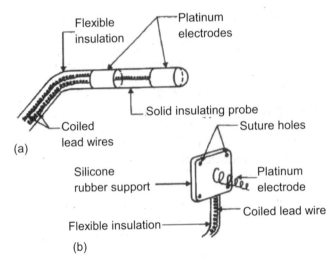

Figure 1. Basic structures of the two types of pacemakers
(a) Bipolar   (b) Unipolar

2.1 Asynchronous Pacemakers

An asynchronous pacemaker gives a fixed heart rate regardless of what is going on in the heart or the rest of the body. It represents the simplest kind of pacemaker. The block diagram of an asynchronous pacemaker is shown in Figure 2. The pulse output circuit of the pacemaker generator produces the actual electric stimulus that is applied to the heart. At each trigger from the timing circuit, the pulse output circuit generates an electric stimulus pulse that has been optimized for stimulating the myocardium through the electrode system that is being applied with the generator. Constant-voltage or constant-current amplitude pulses are the two usual types of stimuli produced by the output circuit. Constant-voltage amplitude pulses are typically in the range of 0.5 to 5.5V with duration of 0.1 to 1.5 ms. Constant-current amplitude pulses are typically in the range of 8 to 10 mA with pulse durations ranging from 1.0 to 1.2 ms. Rates for a synchronous pacemaker range from 70 to 90 beats per minute (bpm).

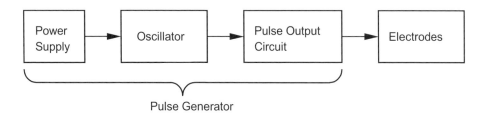

Figure 2. Block diagram of an asynchronous pacemaker

2.2 Synchronous Pacemakers

Many patients can establish a normal cardiac rhythm between periods of heart block. They require cardiac pacing only intermittently. For these patients, continuous stimulating to the ventricles can even result in series complications. Synchronous pacemakers are used to solve these cases. There are two general forms of synchronous pacemakers: the demand and the atrial-synchronous.

2.2.1. Demand Pacemakers.

The block diagram of a demand pacemaker is shown in Figure 3. Except for those elements of the asynchronous pacemaker, it has a feedback loop. A fixed rate, usually 60 to 80 bpm, is set to run for the timing circuit. The feedback signal resets the timing circuit after each stimulus that awaits the appropriate interval to provide the next stimulus. A typical waveform of the electrocardiogram (ECG) is shown in Figure 4 [6]. If a natural beat occurs in the ventricle during this interval, the feedback circuit detects the QRS complex of the ECG signal from the electrodes and amplifies it to reset the timing circuit. If the heart spontaneously beats again before the stimulus of the pacemaker is produced, the timing circuit is again reset and the process repeats itself. Consequently, the heart operates under its own pacing control and the pacemaker is inhibited as in a standby mode. To achieve this, the heart's conduction should operate normally and the natural heart rate should be greater than the rate of the pacemaker. On the other hand, if temporary heart blocks occur, the pacemaker takes over and stimulates the heart at the fixed rate of the timing circuit, which acts as an asynchronous pacemaker.

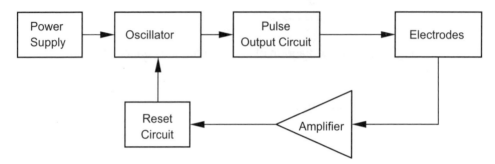

Figure 3.　Block diagram of a demand pacemaker

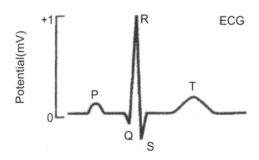

Figure 4.　A typical waveform of the electrocardiogram

2.2.2.　Atrial-synchronous Pacemakers.

　　The heart's physiological pacemaker, located at the SA node (sinoatrial node), initiates the cardiac cycle by stimulating the atria to contract and then provide a stimulus to the AV node (atrioventricular node) that, after appropriate delay for blood to flow from the atria to the ventricle, stimulates the ventricles [6]. If the SA node is able to stimulate the atria, the electric signal corresponding to atrial contraction (the P wave of the ECG) can be detected by an electrode implanted in the atrium and used to trigger the pacemaker in the same way that it triggers the AV node. The block diagram of an atrial-synchronous pacemaker is shown in Figure 5.

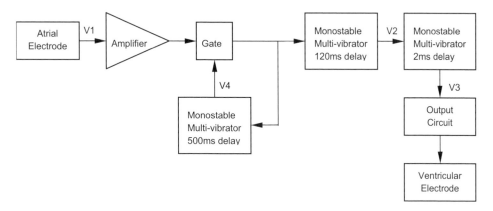

Figure 5.   Block diagram of an atrial-synchronous pacemaker

The waveforms of the signals appearing in the atrial-synchronous pacemaker are shown in Figure 6 [5]. The atrial signal (V_1) detected by the atrial electrode is amplified and passed through a gate to a monostable multi-vibrator giving a pulse (V_2) of 120ms duration, the approximate delay of the AV node. The V_1 signal triggers another monostable multi-vibrator giving a pulse (V_4) of 500ms duration at the same time. The V_4 enables the gate to block any signals from the atrial electrode for a period of 500ms following contraction, i.e. the pacemaker is refractory to any additional stimulation for 500ms after atria contracts. The falling edge of V_2 triggers a monostable multi-vibrator of 2ms duration (V_3). V_2 acts as a delay that let V_3 be produced 120ms after the atrial contraction. Then V_3 triggers an output circuit that applies the stimulus to appropriate ventricular electrode.

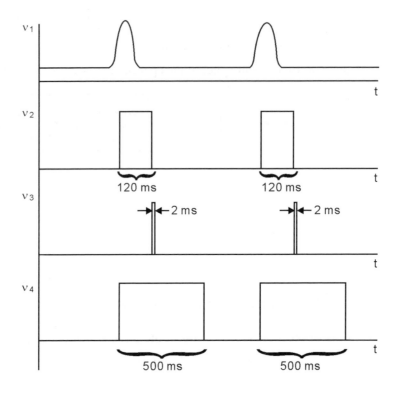

Figure 6. Waveforms of the voltages of an atrial-synchronous pacemaker

Since an atrial-synchronous pacemaker is initiated by the atrial contraction signal, i.e. V_1 detected by the atrial electrode, the pacemaker quits its function if the atrial stimulus is lost. This drawback can be mended by appending a demand pacemaker to an atrial-synchronous pacemaker, namely combined synchronous pacemaker, as shown in Figure 7. The atrial stimulus triggers the reset circuit to disable the oscillator so that it acts as an atrial-synchronous pacemaker. The oscillator takes over and controls the output circuit so that it acts as a demand pacemaker in case the atrial stimulus is absent.

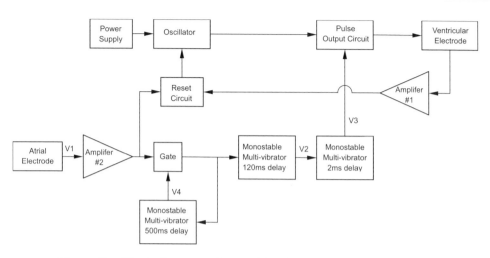

Figure 7.　Block diagram of a combined synchronous pacemaker

2.2.3.　Rate-responsive Pacemakers.

Although the synchronous pacemakers can meet some of the physiological demands for variation in heart rate and cardiac output, these devices still do not replicate the function of the heart in a physiologically intact individual. The demands of a body during stressful activities such as exercise or nervous situation cannot be fully met by these pacemakers. The heart rate should be varied with different demands. To achieve this function, a sensing element should be adopted to convert a physiological variable in the patient to an electrical signal so as to trigger the pacemaker according to the actual demand. The block diagram of this type of pacemaker, namely rate-responsive pacemaker, is shown in Figure 8. The pulse rate of the pacemaker is controlled on the basis of the sensed physiological variable. The controller is able to determine whether an artificial pacing is required and to keep the pacemaker at a dormant state when the patient's natural pacing system is functional. The sensor can be located within the pacemaker or at some other points within the body of the patient [6].

Many different physiological variables have been used to control rate-responsive pacemakers. Table 1 lists some of these variables and the corresponding sensors in an implanted system [7]. Different variable requires different control algorithm for the control circuit. The rest of a rate-responsive pacemaker is the same as an asynchronous one that is aforementioned.

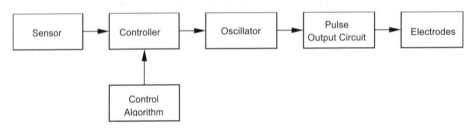

Figure 8.   Block diagram of a rate-responsive pacemaker

Table 1.   Physiological variables and the corresponding sensors for rate-responsive pacemakers

Physiological Variable	Sensor
Right-ventricle blood temperature	Thermistor
ECG stimulus-to-T-wave interval	ECG electrode
ECG R-wave area	ECG electrode
Blood pH	Electrochemical pH electrode
Rate of change of right-ventricle pressure	Semiconductor strain-gage
Venous blood oxygen saturation	Optical oximeter
Intracardiac volume changes	Electric-impedence plethysmography (intracardiac)
Respiratory rate and/or volume	Thoracic electric-impedence plethysmography
Body vibration	Accelerometer

3. PETRI NETS

3.1 Introduction

A Petri net is a general-purpose graphical tool for describing relations existing between conditions and events [8]. The basic symbols of Petri nets include [9]:

○ : Place, drawn as a circle, denotes event

── : Immediate transition, drawn as a thin bar, denotes event transfer with no delay time

━━ : Timed transition, drawn as a thick bar, denotes event transfer with a period of delay time

↑ : Arc, drawn as an arrow, between places and transitions

● : Token, drawn as a dot, contained in places, denotes the data

○ : Inhibitor arc, drawn as a line with a circle end, between places and transitions

Places contain dots, the representation of tokens, being the specific marking of a Petri net [10]. The transition is said to fire, if input places satisfy an enabling condition. Transition firing will remove one token from all of its input places and put one token into all of its output places [11]. Basic structures of logic relations for Petri nets are listed in Figure 9, where P, Q, and R are different events. There are two types of input places for the transition in Figure 9; namely, specified and conditional [12]. The former has a single output arc whereas the latter has multiples. Tokens in a specified-type place have only one outgoing destination, i.e. if the input place(s) holds a token then the transition fires and gives the output place(s) a token. However, tokens in the conditional-type place have more than one outgoing paths, which may lead the system to different situations. For the 'TRANSFER OR' Petri net in Figure 9, whether Q or R takes over a token from P depends on which output-transition of P is fired earlier.

Logic relation	TRANSFER	AND	OR	TRANSFER AND	TRANSFER OR	INHIBITION
Description	If P then Q	If P AND Q then R	If P OR Q then R	If P then Q AND R	If P then Q OR R	If P AND Q' then R
Boolean function	Q=P	R=P*Q	R=P+Q	Q=R=P	Q+R=P	R=P*Q'
Petri nets	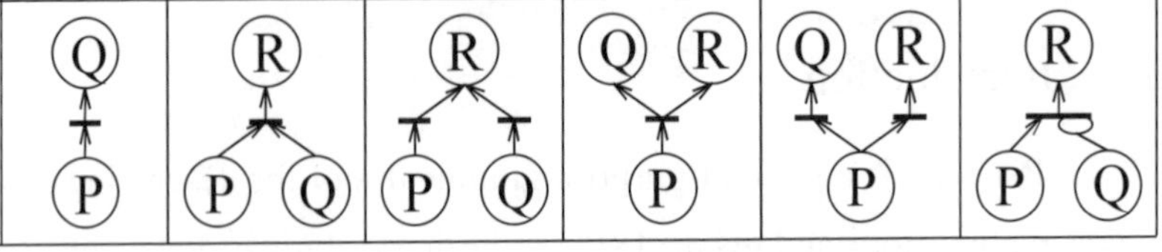					

Fig 9. Basic structures of logic relations for Petri nets

There are three types of transitions that are classified based on time [8]. Transitions with no time delay are called immediate transitions, while those that need a certain constant period of time for transition are called timed transitions. The third type is called a stochastic transition and is used for modeling a process with random time. Hence the Petri net is a powerful tool for modeling various systems. The Petri net for describing the operation of a combined asynchronous pacemaker, i.e. Figure 7, is shown in Figure 10. All the transitions are immediate transitions except T5, T8, T17, and T18. The timed-transition T5 represents the period time of the oscillator while T8, T17, and T18 represents delay-time of 120ms, 2ms, and 500ms, respectively. Buffer places B1 and B2, together with transitions T17 and T9 form a timer, which controls the signal V_3 to trigger the output circuit for 2ms. Similarly, B3, B4, T18, and T6 constitute a timer that activate signal V_4 for 500ms.

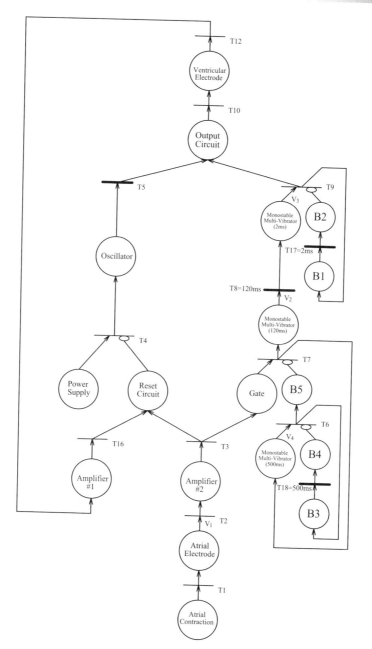

Figure 10.　Petrinet for describing the operation of a combined synchronous pacemaker

3.2 Implementation of Petri Nets

System can be modeled into a Petri net to express not only static behavior such as logical relations between components of the system, but also dynamic behaviors such as operating sequence or failure occurrence of the system. Because Petri nets are state machines [13], it is feasible to realize Petri nets to perform those capabilities. Hardware implementation of Petri nets actualizes state machines that are converted from Petri nets to logic circuits. Petri nets can be implemented as application-specific integrated circuits (ASICs), designed to perform specific functions without user intervention. Mainly because of the programmable capability, field programmable gate arrays (FPGAs) are suitable for hardware implementation of Petri nets. Reference [14] employs a Xilinx FPGA as the design tool to implement Petri nets. The basic symbols for Petri nets are converted to logic circuits as follows. Corresponding circuits are shown in Figure 11 [14].

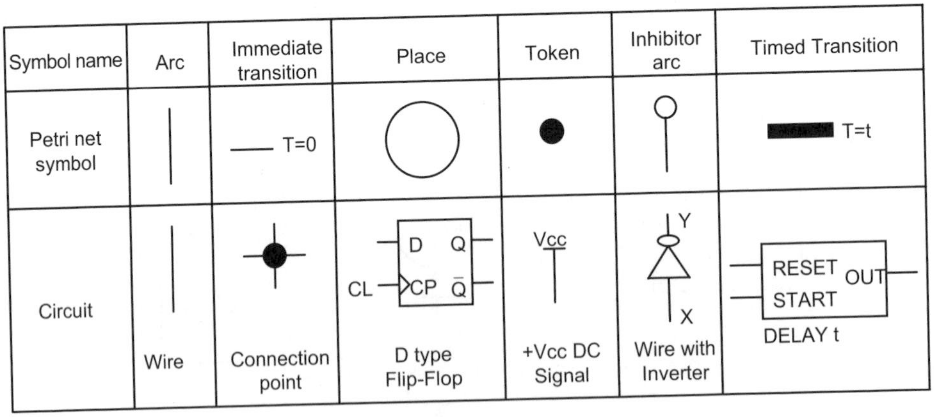

Symbol name	Arc	Immediate transition	Place	Token	Inhibitor arc	Timed Transition
Petri net symbol		—— T=0	◯	●		▬ T=t
Circuit			D type Flip-Flop	Vcc	Y ... X	RESET OUT START DELAY t
	Wire	Connection point	D type Flip-Flop	+Vcc DC Signal	Wire with Inverter	DELAY t

Figure 11. Corresponding circuits for basic Petri net symbols

1. *Place.* A place can be converted to a D type flip-flop, which represents the associated event occurrence modeled by a token by output high Q. Q is high if D is high at the rising edge of the clock pulses.

2.   *Immediate transition.* Since an immediate transition denotes event transfer with no delay time, therefore, a connection point represents it.

3.   *Timed Transition.* It denotes event transfer with delay time t. As shown in Figure 11, it is implemented by a timer with delay time t and start-reset functions. The timer output becomes high at t time later than the arrival of a logic high signal at the timer input.

4.   *Arc.* Arcs model connection wires between components.

5.   *Token.* In logic circuits, a token can be represented as a logic high signal.

6.   *Inhibitor arc.* An inhibitor arc can be converted to a connection wire with an inverter. It inverts the relation between input (X) and output (Y).

4.   FAILURE DIAGNOSIS

A marking of a Petri net is defined as the total number of tokens at each place, denoted by a column vector M. Thus vector $M_k = [m(P_1), m(P_2), ... m(P_m)]^T$ represents that token numbers of places P_1, P_2, ... P_m at state k are $m(P_1)$, $m(P_2)$, ... $m(P_m)$, respectively [12]. After adding several checkpoints into Figure 10, the working status of a combined synchronous pacemaker can be clearly visualized on the marking of the checkpoints [12]. Figure 12 shows the resultant Petri net for failure diagnosis. The added places (shadowed portion in Figure 12), i.e. CP1, CP2, ..., and CP12, are defined as follows:

CP1:   Checkpoint 1, m(CP1)=1 (0) represents that the power-supply is functioning (not functioning).

CP2:   Checkpoint 2, m(CP2)=1 (0) represents that the atrial-electrode is functioning (not functioning).

CP3:   Checkpoint 3, m(CP3)=1 (0) represents that the amplifier#2 is functioning (not functioning).

CP4:   Checkpoint 4, m(CP4)=1 (0) represents that the reset-circuit is functioning (not functioning).

CP5: Checkpoint 5, m(CP5)=1 (0) represents that the oscillator is functioning (not functioning).

CP6: Checkpoint 6, m(CP6)=1 (0) represents that the 500ms-delay-vibrator is functioning (not functioning).

CP7: Checkpoint 7, m(CP7)=1 (0) represents that the gate is at a closed state (an open state).

CP8: Checkpoint 8, m(CP8)=1 (0) represents that the 120ms-delay-vibrator is functioning (not functioning).

CP9: Checkpoint 9, m(CP9)=1 (0) represents that the 2ms-delay-vibrator is functioning (not functioning).

CP10: Checkpoint 10, m(CP10)=1 (0) represents that the output-circuit is functioning (not functioning).

CP11: Checkpoint 11, m(CP11)=1 (0) represents that the ventricular electrode is functioning (not functioning).

CP12: Checkpoint 12, m(CP12)=1 (0) represents that the amplifier#1 is functioning (not functioning).

Besides, five buffer-places, i.e. B6, B7, ..., B10, are employed to complete the Petri net.

In Figure 12, if the atria contracts, the transition T1 fires such that the Atrial-Electrode obtains a token. Immediately, T2 fires and each of Amplifier#2 and CP2 obtains a token. With no time-delay, transition T3 fires to feed each of the Gate, CP3, and the Reset Circuit a token. Consequently, each of CP4 and B7 is fed a token through the firing of T19. Because the place B5 contains no token yet at this moment, thus transition T7 fires such that both the 120ms-delay-vibrator and the 500ms-delay-vibrator are triggered to function at the same time. CP7 acquires a token as well. Afterwards, each of CP6, B8, CP8, and B9 obtains a token. As a result, the feedback signal V_4 inhibits the firing of T7 for 500ms duration, and the 120ms time delay of V_2, i.e. T8, is started to count down. When 120ms after the firing of T22, the 2ms-delay-vibrator is activated for 2ms and each of CP9 and B10 obtains a token at the same time. Consequently, the Output Circuit is triggered to function for 2ms and CP10 obtains a token. Then, the Ventricular Electrode is triggered to function such that the marking of CP11 becomes unity, i.e. the atrial-synchronous pacemaker is functioning. On the left-hand side of Figure 12, however, the firing of transition T4 is inhibited if place "Reset Circuit" contains a token. Therefore, the function of the demand pacemaker is disabled. The Power-Supply supplies power to the whole system and the marking of CP1 indicates the states of the Power-Supply. In case the atrial signal (V_1) is lost, T2 is thus disabled to suspend the function of the atrial-synchronous pacemaker and T4 fires to enable the demand pacemaker.

Let M_{CP} be the marking of the twelve checkpoints, i.e. $M_{CP} = [m(CP1), m(CP2), ..., m(CP12)]^T$, then the health condition of the pacemaker is visualized via M_{CP}. If the atrial stimulus is absent, i.e. the demand pacemaker is functioning, the marking of the twelve checkpoints is

$$M_{CP} = [1, 0, 0, 0, 1, 0, 0, 0, 0, 1, 1, 1]^T.$$

On the contrary, if the atria contracts normally, i.e. the atrial-synchronous pacemaker is functioning and the fixed-rate oscillator is dormant, the marking of the checkpoints becomes

$$M_{CP} = [1, 1, 1, 1, 0, 0, 1, 0, 0, 0, 0, 0]^T \quad (0{\sim}0^+\text{ms}).$$

The sixth entry of M_{CP} is 0 because the 500ms-delay-vibrator is not activated yet at this moment. After the 500ms-delay-vibrator is triggered, the markings of the checkpoints are shown as follows.

$$M_{CP} = [1, 1, 1, 1, 0, 1, 0, 1, 0, 0, 0, 0]^T \quad (0{\sim}120\text{ms}),$$
$$M_{CP} = [1, 1, 1, 1, 0, 1, 0, 0, 1, 1, 1, 1]^T \quad (121{\sim}122\text{ms}),$$
$$M_{CP} = [1, 1, 1, 1, 0, 1, 0, 0, 0, 0, 0, 0]^T \quad (123{\sim}500\text{ms}).$$

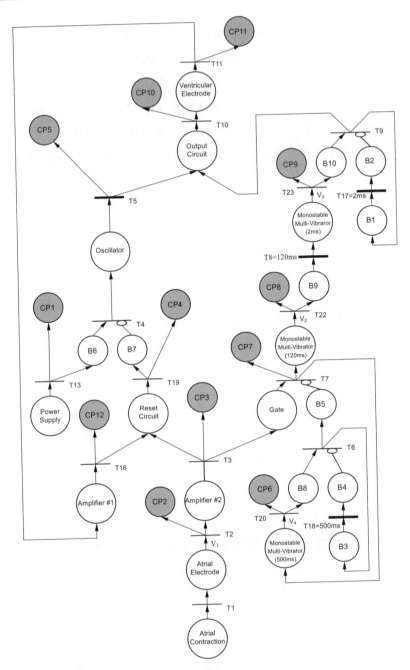

Figure 12.   Petri net for failure analysis of a combined synchronous pacemaker

5.　REMOTE DIAGNOSIS

5.1　Control Chart and Threshold

A failure threshold is a value used to judge if an equipment failure occurs or not. It is prescribed as the measurement value that is taken just prior to or at the time of failure [15]. Life testing is one method to obtain such data, and may be performed by field-engineers or users. Normally, the mean value of a failure-probability function that is established from tests of manufacturers is a theoretic value for the threshold.

Once the threshold has been determined, a margin of safety should be added to account for variations in early failure detection. The safety margin can be determined by the requirement of lead-time for preventive maintenance (PM) or evaluation of the physical properties and actual operating conditions of different systems. The lower the warning value is set, the greater is the assurance that PM will be done prior to failure [15], whereas more labor manpower and cost will be expended. Theoretically, triple the standard deviation is one possible choice in prescribing a warning value [16]. On the basis of failure thresholds and warning values, a control chart can be constructed to conduct limit control, as illustrated in Figure 13 [12]. The lead-time of early detection can be obtained by extrapolating the curve in a control chart with a line slope that is constructed by the last two sampled points on the curve [17]. The lead-time is the period between the time point where the warning value is exceeded and the intersection of the extended line and the time-axis. The lead-time obtained from the control chart is for the action of the PM for the device.

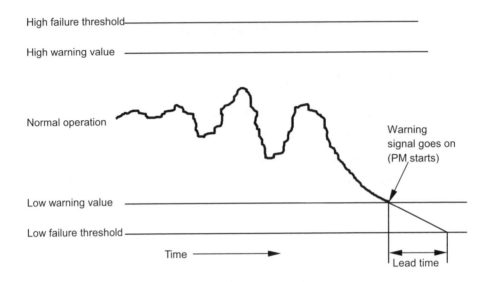

Figure 13. A control chart

Since batteries in implanted pacemakers are the main cause of pacemaker failure, it is worthwhile to replace the batteries before they cannot supply enough power in order to avoid dangerous situations caused by low battery-voltage for the patients. Therefore, a voltage measuring circuit should be integrated with the circuit of a pacemaker to monitor the battery-voltage level. Besides, a comparator is also needed in the circuit to compare the measured battery-voltage level to the prescribed low warning value for the battery so as to generate a warning signal when the battery-voltage level is lower than the prescribed low warning value.

5.2 Remote Mode

Once a cardiac pacemaker is implanted, it becomes difficult to perform trouble-shooting and maintenance on it. Hence, an implanted pacemaker should have high reliability and low requirement of maintenance. However, knowing when and why an implanted pacemaker should be extracted are important to the user of it. To achieve these purposes, wireless communication is one possible solution. Figure 14 shows a Petri net arrangement that realizes these functions. As shown in Figure 14, a low-power transmitter is integrated with the combined synchronous pacemaker that was shown in Figure 12, which is used to transmit the output data of the mixer. The inputs of the mixer are CP2, CP3, ..., CP12, and the battery-voltage level that is measured by a measuring circuit. The twelve data can be modulated as a checking-code to show the working status of the pacemaker. For explaining how the remote mode works, the Petri net in Figure 14 is divided into four portions.

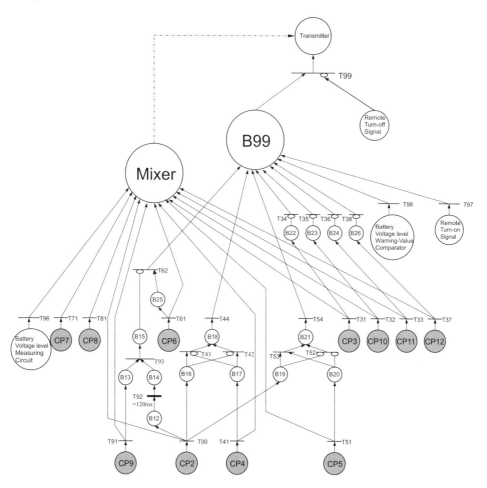

Figure 14.   Petri net for the remote mode of a combined synchronous pacemaker

Portion 1: *Remote-control function of the transmitter.* A remote-controller separately generates a remote turn-on or a remote turn-off signal by depressing the corresponding button on the controller to operate the transmitter. The dot line between the mixer and the transmitter implies that the transmitted signal is fed by the mixer. A receiver located at the outside of the patient's body, which is not shown in Figure 14, displays the checking-code (and sounds a beep if it is necessary) whenever it receives the signal from the transmitter to remind the patient. In Figure 14, the transition T99 fires to function the transmitter if place B99 holds a token and the remote turn-off signal is absent. Either transition T98 or T97 fires can feed B99 a token. In other words, whenever the battery-voltage level is lower than the prescribed warning low value, the transmitter is triggered to function automatically; or whenever it is necessary to read the checking-code, the transmitter is also triggered to function by depressing the turn-on button manually. On the other hand, the transition T99 is inhibited to disable the transmitter whenever the remote turn-off signal is present.

Portion 2: *Common parts for the demand pacemaker and the atrial-synchronous pacemaker.* The B99 is fed a token if the marking of either one of CP3, CP10, CP11, or CP12 is zero, i.e. one of the failure represented by CP3, CP10, CP11, or CP12 occurs, the transmitter is triggered to transmit the checking-code automatically.

Portion 3: *The Demand pacemaker.* Whether the "Reset Circuit" in Figure 10 functions or not, which is represented by the marking of CP4 in Figure 12, relates to the atrial contraction. If the atria contracts, i.e. $m(CP2)=1$, the reset circuit should function to reset the oscillator, i.e. $m(CP4)=1$, such that the atrial-synchronous pacemaker is on duty. This represents the pacemaker at a normal condition. Accordingly, The B99 should not be fed a token to trigger the transmitter, i.e. $m(B99)=0$. On the other hand, both the demand-pacemaker and the atrial-synchronous pacemaker function at the same time if the atria contracts whereas the reset circuit does not function, which is an abnormal condition so that the B99 should be fed a token (i.e. m(B99)=1) to trigger the transmitter. Discussing the rest conditions in a similar manner, the relations among markings of CP2, CP4, and B99 are summarized in Table 2. Similarly, whether the oscillator functions or not, which is represented by the marking of CP5, is also related to the marking of CP2. The relations among markings of CP2, CP5, and B99 are summarized in Table 3. Clearly, the logic relations for Table 2 and Table 3 are exclusive-OR (XOR) and exclusive-not-OR (XNOR), respectively. The Perti nets and the converted circuits for XOR and XNOR are shown in Figure 15 [14].

Table 2.   Truth table for the relations among CP2, CP4, and B99

CP2	CP4	B99
1	1	0
1	0	1
0	0	0
0	1	1

Table 3.   Truth table for the relations among CP2, CP5, and B99

CP2	CP5	B99
1	1	0
1	0	0
0	0	1
0	1	0

Logic relation	XOR	XNOR
Description	If X1 not equals to X2 then Y	If X1 equals to X2 then Y
Booklean function	$Y = X1 \times \overline{X2} + \overline{X1} \times X2$	$Y = X1 \times X2 + \overline{X1} \times \overline{X2}$
Retri nets	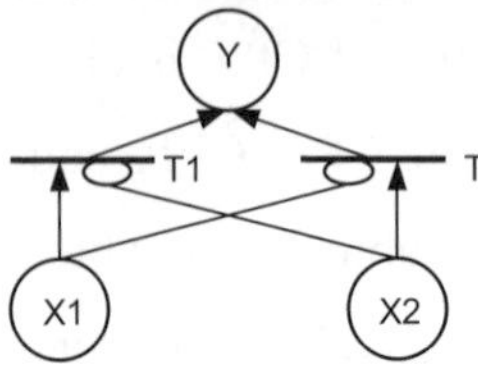	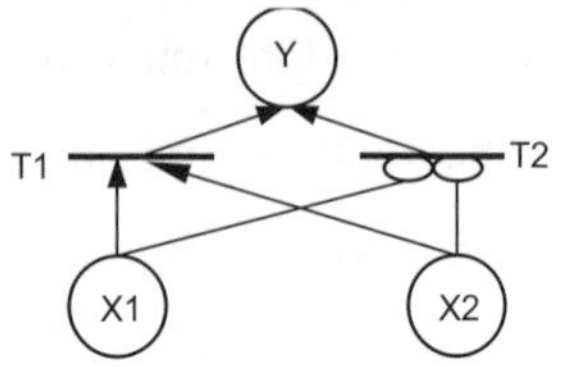
Citcuit	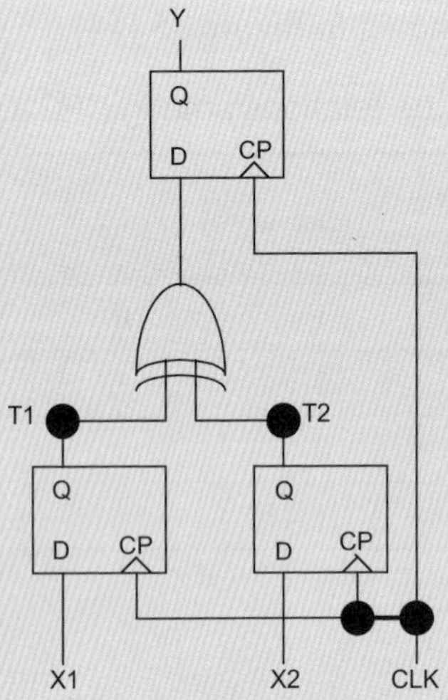	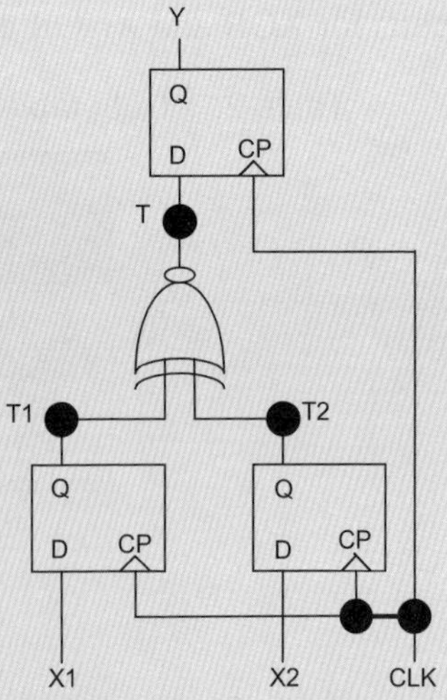

Figure 15.   Structures of XOR and XNOR for Petri nets and corresponding circuits

Portion 4: *The Atrial-synchronous pacemaker.* The atrial-synchronous pacemaker is on duty and works according to the timing sequence as shown in Figure 6 if the atria contracts normally (m(CP2)=1). Transition T00 fires to feed buffer place B12 a token. Since T92 is a timed transition that represents 120ms delay, buffer place B14 obtains a token when 120ms after the atrial contracts. After B14 holds a token, T93 fires to feed B15 a token if CP9 also contains a token. The marking of m(CP9)=1 represents the appearance of signal V_3. In Figure 6, the relations among signals V_1, V_2, and V_3 only last for 500ms that equals the duration of signal V_4. Accordingly, in Figure 14, buffer place B25 holds a token for only 500ms. During these 500ms, T62 is inhibited and does not feed a token to B99 if B15 holds a token. The marking of m(B15)=1 represents the atrial-synchronous pacemaker at a normal condition. On the contrary, T62 fires to function the transmitter if B15 is empty during the 500ms, which represents the atrial-synchronous pacemaker at an abnormal condition. The buffer place B25 acts as a 500ms-timer so that transition T62 has the chance to fire only within the 500ms when B25 holds a token. T62 does not fire no matter B15 holds a token or not during the time other than the 500ms.

In summary, the remote mode for combined synchronous pacemakers has the following functions:

1. The working status of the pacemaker is shown by the checking-code that is a result of mixing twelve measured data points of the pacemaker.
2. A low-power transmitter can transmit the output of the mixer, i.e. the checking-code, from an implanted pacemaker to the outside of the patient's body to remind the patient (with a beep sound if it is necessary).
3. The transmitter can be turned on manually by depressing the turn-on button on the controller at any time.
4. The transmitter can be turned on automatically if the pacemaker is at an abnormal condition.
5. The checking-code shows where the failure is located if the pacemaker is out of order.
6. The present voltage level of the battery in the implanted pacemaker is also shown in the checking-code.
7. The transmitter can be turned off manually by depressing the turn-off button on the controller at any time.

6. CONCLUSIONS

Cardiac pacemakers are very helpful to patients with heart block problems. The control algorithm directly affects the adaptability of a pacemaker to a human body. Consequently, research into new control algorithms for constructing more adaptive and more reliable pacemakers is under way. The Petri net is a powerful graphical tool for modeling a dynamic system such as a combined synchronous pacemaker, which helps the design, failure diagnosis, and research of control algorithms of a cardiac pacemaker. This study demonstrated the modeling and failure diagnosis for the normal mode and the remote mode of a combined synchronous pacemaker by the Petri net approach. The operational status of the pacemaker is clearly visible from the Petri net model and the health condition is clear at a glance by the checking-code of the pacemaker.

REFERENCES

1. Dhillon BS. *Mechanical reliability: theory, models and applications*; AIAA Education Series: Washington DC, 1988.

2. Chiou SB. Failure analysis in reliability engineering using Petri nets. *M.S.-thesis*; National Chiao Tung University, Taiwan, Republic of China, 1995.

3. O'Connor PDT. *Practical reliability engineering*; John Wiley: Chichester, 2002.

4. Ereau JF, Saleman M. Modeling and simulation of a satellite constellation based on Petri nets. *Proceedings of the Annual Reliability and Maintainability Symposium IEEE*, 1996; 66-72.

5. Webster JG. *Medical instrumentation application and design*; John Wiley & Sons: New York, 1995.

6. Furman S, Escher DJW. *Principles and techniques of cardiac pacing*; Harper & Row: New York, 1970.

7. Smith HJ, Fearnot NE. Concepts of rate-responsive pacing. *IEEE Eng. Med. Biol. Mag.* 1990; **9**(2):32-35.

8. David R, Alla H. Petri nets for modeling of dynamic systems-A survey. *Automatica* 1994; **30**(2):175-202.

9. Hura GS, Atwood JW. The use of Petri nets to analyze coherent fault trees. *IEEE Transactions on Reliability* 1988; **37**(5):469-474.

10. Schneeweiss WG. *Petri net for reliability modeling*; Lilole-Verlag, 1999.

11. Yang SK, Liu TS. Failure analysis for an airbag inflator by Petri nets. *Quality and Reliability Engineering International* 1997; **13**:139-151.

12. Yang SK, Liu TS. A Petri net approach to early failure detection and isolation for preventive maintenance. *Quality and Reliability Engineering International* 1998; **14**:319-330.

13. Shaw AW. *Logic circuit design*; Saunder College Publishing: Fort Worth, 1993.

14. Yang SK, Liu TS. Implementation of Petri nets using a field-programmable gate array. *Quality and Reliability Engineering International* 2000; 16:99-116.

15. Patton Jr. JD. *Preventive maintenance*; Instrument Society of America: New York, 1983.

16. Rao SS. Reliability-based design; McGraw-Hill: New York, 1992.

17. Kumamaru T, Utsunomiya T, Yamada Y. A fault diagnosis system for district heating and cooling facilities. *Proceedings of the IECON'91, International Conference on Industrial Electronics, Control, and Instrumentation*, 1991; 131-136.

附錄 C

發電廠中根據狀態而實施維護之研究

摘　要

本文利用混合式裴氏網路建模法，另配合參數趨勢及故障樹分析，提出了一具有多項功能之裴氏網路規劃。該規劃之功能包括：故障警告、早期失效偵測、故障隔離、事件計數、系統狀態描述、以及自動關機或自動調節。這些功能對於系統狀態正常與否之監視及預防維護策略之安排均非常有用。本文並以傳統火力發電系統為例，說明該規劃之功能及特性。

＊關鍵詞：混合裴氏網路、早期失效偵測及隔離、預防維護、故障樹分析、趨勢圖

介　紹

維護的執行可分為根據狀態的維護(Condition-based maintenance)和根據時間的維護(Time-based maintenance)兩種。如果按照故障發生之前或後，也可將維護分為預防維護和矯正維護兩大類(Patton，1983)。預防維護是在系統發生失效之前，對系統進行檢測，更換零組件、潤滑及調整；旨在減少系統發生失效所帶來的損失，使系統可靠度得以提升，並且透過對系統老化現象的補救而延長系統的壽命(蔡明三，1982)。

我國的發電廠一向採取時間為基礎的維護策略，也就是根據設備廠商的建議或者現場人員的維修經驗，決定實施維護的時間。因為這些根據係統計平均值，所以實際上實施維護的時候，設備或元件可能仍處於良好狀態，比真正需要實施維護的時間提早得多，以致形成人力、物力與可發電力的無謂損失。反之，也可能比真正需要實施維護的時間延遲，以致釀成故障停機的損失。如何訂定合理的維護策略，確實是發電廠應該重視的課題。欠缺妥善的維護策略，會使系統有時處於不安全的狀態；反之，實施妥善的維護策略能夠使系統隨時

保持在滿意或良好的狀態。本研究旨在利用混合式裴氏網路及參數趨勢的方法，針對我國發電廠的機械設備，憑藉監視系統的資料，發展出根據設備的使用狀態而執行的維護策略與方法，使得我國發電廠的維護計畫能夠從「根據時間的維護」邁向「根據狀態的維護」。

門檻值及警告值

誤差(Error)、失效(Failure)、故障(Fault)三者間之關係示於圖一，其定義分別如下(IEC，1990)：

1. 誤差是指計算值、觀察值、量測值或條件與眞值、指定值、理論正確值或條件間之差異。
2. 失效是一事件，指要求的功能終結或誤差已達所能接受的界限。
3. 故障是一狀態，指要求的功能無法表現，不包括正在執行預防維護或其他計劃中的行動，或因缺少外部資源而導致的失能狀態。

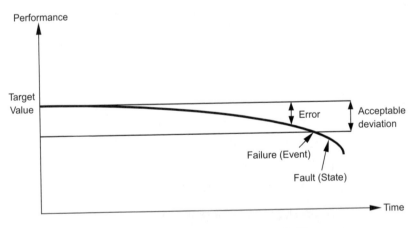

圖一　誤差(Error)、失效(Failure)、故障(Fault)

根據上述定義可知，誤差並非失效而故障是失效所導致的狀態。誤差有時也被稱爲初始失效(Rausand，1996)，所以預防維護實施的時間是在當系統仍處於誤差狀態，也就是介於圖一中之可接受變異區間(Acceptable deviation)，而早於失效發生之時。因此，經由預防維護技術，失效可以提早被偵知。

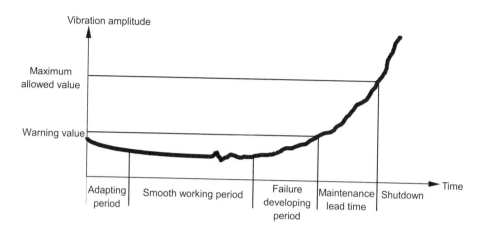

圖二　旋轉機械的趨勢圖

　　依時間記錄設備之性能表現的圖形稱爲趨勢圖(Barna，1992)。圖二是旋轉機械之趨勢圖，即爲一例。趨勢圖通常是由製造商在實驗室裡評估產品品質、可靠度、維護能力、及維護程序時所建立。門檻值是一數值，爲判定一設備是否失效的依據。該值是指失效即將發生或剛發生時的性能量測值，也就是圖二中的最大容許值(Maximum allowed value)。壽命試驗是獲取門檻值的方法之一，該值亦可由現場工程師或使用者依實務經驗來決定。門檻值一經決定之後，另需一安全範圍作爲變化容許量以執行早期故障偵測。圖二中之警告值(Warning value)即爲此範圍之邊界。安全範圍可依預防維護所要求的前置時間來決定，或可隨不同系統實際操作狀況及物理特性而訂定。警告值設得愈低，可保障執行預防維護的時間愈早於失效發生的時間(Patton，1983)，然如此會耗費較多人力及成本。失效分配函數之標準差的三倍，是決定警告值可行的選擇之一(Rao，1992)。依據失效門檻值及警告值即可建立一控制圖(Patton，1983)以執行性能極限控制，如圖三所示。

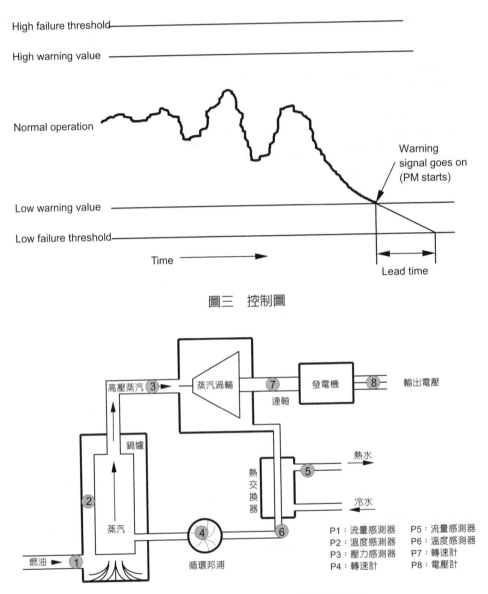

High failure threshold

High warning value

Normal operation

Warning signal goes on (PM starts)

Low warning value

Low failure threshold

Time

Lead time

圖三　控制圖

高壓蒸汽 ③　蒸汽渦輪　⑦　發電機　⑧　輸出電壓

連軸

鍋爐

熱交換器

熱水

⑤

冷水

②

蒸汽

④

循環邦浦

⑥

燃油　①

P1：流量感測器　　P5：流量感測器
P2：溫度感測器　　P6：溫度感測器
P3：壓力感測器　　P7：轉速計
P4：轉速計　　　　P8：電壓計

圖四　配置資料擷取系統的火力發電系統

　　失效偵測是指將實際值與公稱值作比較，而故障隔離是將實際值與故障值作比較(Baccigalupi，1997)。所以，為執行預防維護必須建立一套資料收集系統，以擷取量測點的實際值。除了作比較用途之外，量測值亦可儲存建立成一

資料庫，以修正之前所訂定之門檻值及警告值。某些設備的性能表現是隨外在條件而變化的，例如發電機在一天當中之輸出電流係隨負載而變。所以門檻值及警告值可以依事先規劃好的時序來變動，以達到適應調整之功能。

本文以一傳統火力發電系統為例子，說明如何以本研究所提出之方法來達到早期失效偵測及故障隔離的目的，該系統示於圖四。為了建立該系統之失效分析裴氏網路，有八個感測器經選擇後分別安裝在不同的測試點上。感測器的型式、安裝位置、及相對應之感測信號，均描述於圖四之中。

裴氏網路

裴氏網路(Petri nets)是一種通用型的數學工具，用以描述事件與條件間的關係(David，1994)。裴氏網路的基本符號包括(Hura，1988)：

○ ： 位置，以一圓圈表示，代表事件

── ： 立即變遷，以一細棒表示，代表事件轉移時無時間延遲

━━ ： 時延變遷，以一粗棒表示，代表事件轉移時有時間延遲

↑ ： 弧，以一箭頭表示，介於位置與變遷中間

● ： 標記，以一圓點表示，含於位置中，代表資料

♀ ： 禁止弧，以一圓圈與一直線表示，介於位置與變遷中間

Logic relation	TRANSFER	AND	OR	TRANSFER AND	TRANSFER OR	INHIBITION
Description	If P then Q	If P AND Q then R	If P OR Q then R	If P then Q AND R	If P then Q OR R	If P AND Q' then R
Boolean function	Q=P	R=P*Q	R=P+Q	Q=R=P	Q+R=P	R=P*Q'
Petri nets	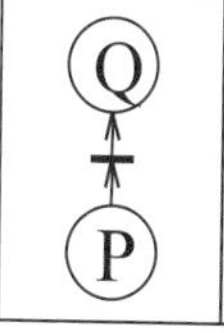	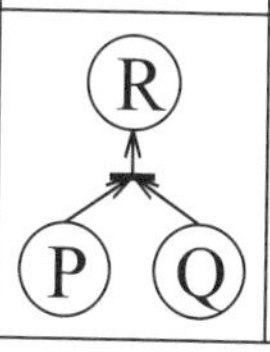	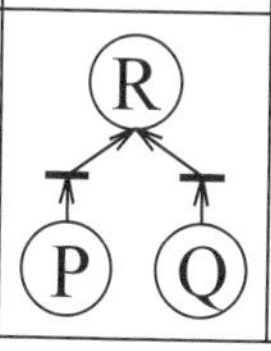	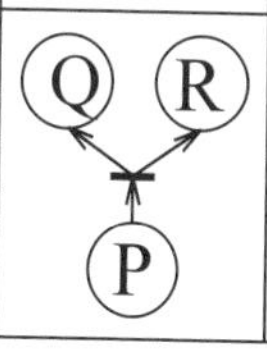	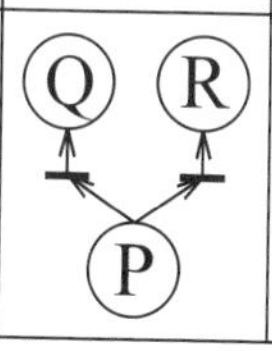	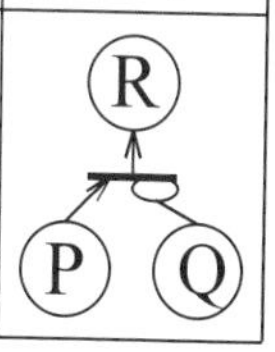

圖五　事件間邏輯關係基本結構的裴氏網路

如果輸入位置滿足了使能條件，則變遷可發射。變遷發射會使變遷之每一個輸入位置移出一個標記，並放入一個標記至其每一個輸出位置中(Schneeweiss，1995)。以裴氏網路表示事件間邏輯關係的基本結構示於圖五。

圖中變遷之輸入位置可分爲既定型和條件型兩類。既定型只有一個輸出弧，而條件型之輸出弧則有多個。在既定型位置中之標記僅有一個去處，也就是說如果該位置擁有一個標記，在變遷發射後，將使得該變遷之每一輸出位置均獲得一個標記。然而在條件型中的標記因有多個出路，故會依條件將系統導引至不同的狀況。以圖五中「TRANSFER OR」的裴氏網路爲例，是 Q 或是 R 位置會由 P 位置取得一個標記，將視例如：機率、外界動作、或內部狀況等條件而定。另依耗時情形可將變遷分爲三類：轉移過程沒有任何時間延遲的變遷稱爲立即變遷，而延遲時間爲常數者稱爲時延變遷。第三類稱爲隨機變遷，是用來模擬時間延遲爲隨機的程序。裴氏網路對於失效分析及預防維護甚爲有用(Yang，1997；1998)。圖六爲傳統火力發電系統之失效分析裴氏網路。

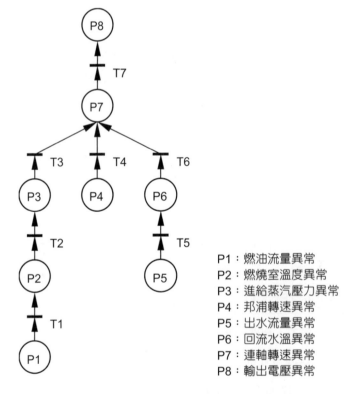

P1：燃油流量異常
P2：燃燒室溫度異常
P3：進給蒸汽壓力異常
P4：邦浦轉速異常
P5：出水流量異常
P6：回流水溫異常
P7：連軸轉速異常
P8：輸出電壓異常

圖六　火力發電系統的失效分析裴氏網路

早期失效偵測與隔離規劃

本研究提出了一早期失效偵測與隔離規劃(EFDIA)，它是一混合式裴氏網路，包括了一般型、禁止弧型、及時延型三種裴氏網路。在圖六中每一位置，也就是圖四中的每一量測點，均配置有一個 EFDIA 用以執行故障警告、早期失效偵測、故障隔離、事件計數、系統狀態描述、及自動關機或自動調節等功能。EFDIA 如圖七所示，其中符號定義為:

1. n ：感測點總數。

2. i ：序號；$1 \leq i \leq n$

3. $M(P)_k$ ：位置 P 在第 k 狀態時的標識(Marking)；代表位置 P 在第 k 狀態時標記的數目，k = 1, 2, 3, …。

4. P_i ：裴氏網路的第 i 個位置；如果 P_i 發生失效則 $M(P_i) = 1$。

5. T_i ：裴氏網路的第 i 個變遷；代表變遷過程所持續的時間。

6. S_i ：第 i 個位置的感測信號；如果本信號超過了規定的警告值，也就是一異常狀況(Error，誤差)發生則 S_i 產生一個標記使得 $M(S_i) = 1$。

7. T_{iE} ：第 i 個位置的誤差變遷；立即變遷。

8. T_{iL} ：第 i 個位置的誤差次數記錄變遷；亦為立即變遷。

9. T_{iM} ：第 i 個位置的維護變遷；代表對 P_i 實施預防維護由開始至結束所需的過渡時間，為時延變遷。

10. T_{iP} ：第 i 個位置的處理變遷；立即變遷。

11. T_{iR} ：P^R_i 位置的重置變遷；立即變遷。

12. T_{iS} ：第 i 個位置的感測變遷；立即變遷。

13. T_{iT} ：第 i 個位置的轉移變遷；立即變遷。

14. T_{iU} ：第 i 個位置的無處理變遷；代表由第 i 個警告信號出現至第 i 個位置的失效發生所需的過渡時間，為時延變遷。

15. T_{iW} ：下一層的 P^w 警告次數記錄變遷；為立即變遷。

16. P^A_i ：第 i 個位置採取預防維護的位置；如果第 i 個位置採取預防維護則 P^A_i 產生一個標記使得 $M(P^A_i) = 1$。

17. P^{Bj}_i ：第 i 個位置的第 j 個緩衝位置；可暫存標記，$j = 1, ..., x$; x 是第 i 個位置輸入弧的數目。

18. P^E_i ：第 i 個位置的誤差指示位置；如果 $M(S_i) = 1$ 的情形是由 P_i 本身而不是由下一層位置所造成，則經由 T_{iE} 的發射使得 $M(P^E_i) = 1$(作為故障隔離之用)。

19. P^F_i ：第 i 個位置的失效計數位置；$M(P^F_i)$ 代表第 i 個位置的失效次數。第 i 個位置每失效一次則 $M(P^F_i)$ 增加 1。

20. P^L_i ：第 i 個位置的誤差計數位置；$M(P^L_i)$ 代表第 i 個位置的誤差次數。第 i 個位置每出現誤差一次則 $M(P^L_i)$ 增加 1。

21. P^M_i ：第 i 個位置的維護計數位置；$M(P^M_i)$ 代表第 i 個位置的維護次數。若 $M(S_i) = 1$ 的狀況經維護則 $M(P^M_i)$ 加 1。

22. P^P_i ：第 i 個位置的正在維護狀態位置；代表 P_i 正處在維護狀態。

23. P^R_i ：第 i 個位置的重置計數位置；$M(P^R_i)$ 代表第 i 個位置由下一層位置所引起的警告次數，也就是第 i 個 RESET R 的重置次數；第 i 個 RESET R 每觸動一次則 $M(P^R_i)$ 增加 1。

24. P^T_i ：第 i 個位置的過渡狀態位置；代表介於 S_i 與 P_i 間的過渡狀態，其持續時間為 T_{iS} 加 T_{iT}。該狀態是在無 EFDIA 時 S_i 至 P_i 間原來的路徑。

25. P^U_i ：第 i 個位置的無處理位置；$M(P^U_i) = 1$ 代表第 i 個位置發生誤差但未矯正。

26. P^W_i ：第 i 個位置的警告計數位置；$M(P^W_i)$ 代表第 i 個位置發生警告的次數，不論該警告信號肇因何處；第 i 個 RESET W 每觸動一次則 $M(P^W_i)$ 增加 1。

27. ith RESET E：第 i 個位置的重置 E 位置；代表對 P^E_i 的重置信號;每觸動一次則產生一個標記。

28. ith RESET R：第 i 個位置的重置 R 位置；代表對 P^R_i 的重置信號;每觸動一次則產生一個標記。

29. ith RESET W：第 i 個位置的重置 W 位置；代表對 $P^B{}_i$ 的重置信號；每觸動一次則產生一個標記；此重置 W 位置的數目須等於變遷 T_{iE} 中禁止弧的數目。

30. ASFM：自動關機或回授機構；例空調或通風系統是防止超溫的回授機構。

31. ith WARNING SIGNAL：第 i 個位置的警告位置。

32. CLOCK：內建的時鐘，用來記錄事件發生的時間。

    EFDIA 的動作步驟綜整於圖八的流程圖中。EFDIA 是由許多小的規劃所構成，每個小規劃均有其特定功能，說明如下：

1. 重置。由 ith RESET E、$P^E{}_i$、及 T_{iL} 三者形成了邏輯 AND 的關係。當 $P^E{}_i$ 擁有一個標記且須要被重置時，可經由觸發 ith RESET E，由 ith RESET E 產生一個標記，使得 T_{iL} 發射，從而將標記由 $P^E{}_i$ 移至 $P^L{}_i$。同樣地，ith RESET R 和 ith RESET W 是分別被設計用來重置 $P^{B1}{}_i$ 和 $P^{B2}{}_i$ 的。

2. 禁止。T_{iE} 和 T_{iU} 兩者均連有一禁止弧，此規劃是用來禁止連有禁止弧的位置(即 $P^{B2}{}_i$ 和 $P^A{}_i$)擁有標記時，其所分別對應之變遷(即 T_{iE} 和 T_{iU})發射之用。

3. 條件型位置。只要有兩個以上輸出弧的位置即稱為條件型位置。條件型位置中的標記會經由其所有輸出弧中最早致能的變遷而移至該致能變遷的輸出位置。$P^A{}_i$、$P^{B1}{}_i$、$P^{B2}{}_i$ 以及 ith WARNING SIGNAL 為條件型位置。

4. 計數器。$P^F{}_i$、$P^L{}_i$、$P^M{}_i$、$P^R{}_i$、及 $P^w{}_i$ 為計數器，分別累計相對應事件發生的次數。

5. 事件旗號。有許多位置被設計成事件旗號，當其中任一位置的標記為 1 時，即表示相對應之事件已發生。

   (1) P_i 表示第 i 個位置失效。

   (2) $P^E{}_i$ 表示誤差位於第 i 個位置。

   (3) $P^T{}_i$ 表示第 i 個位置正處於過渡狀態。

   (4) ith WARNING SIGNAL 表示第 i 個監測信號已達到警告值。

   (5) ASFM 表示自動關機功能或回授機構已經啟動。

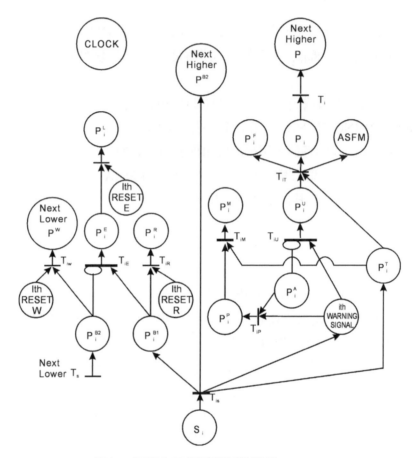

圖七　早期失效偵測與隔離規劃(EFDIA)

EFDIA 所具有的功能及從其中推導所得之之不變量，討論於後。

1. 功能

(1) 警告：任一位置發生監測值超越警告值的狀況時，該位置之 WARNING SIGNAL 即被觸動，亦即於該位置發出警訊。

(2) 早期失效偵測：EFDIA 是在監測信號達到預設警告值的時候發出警訊，而不是在達到失效門檻值的時候。這表示異常狀況是在失效發生之前即被偵知。如圖三所示，將控制圖中曲線抵達警告值前之最後兩

點連成直線(Kumamaru，1991)，由此線與門檻值相交之點與曲線抵達警告值之點間所夾的時間，即為早期偵知的領先時間。

(3) 故障隔離：系統的任何地方均可能是造成功能異常的原因所在。既然功能異常的原因已被故障樹中的邏輯關係所規範，這些原因即可被變遷 T_{iE} 所隔離，並由事件旗號 P^E_i 所指示。只要 $M(P^E_i)$ 等於 1，即代表該誤差是位於第 i 個位置，否則既使第 i 個警告信號出現，仍可知此誤差係起因於下層位置，而不是第 i 個位置。

(4) 事件計數：所有在 EFDIA 中的計數器均可累計相對應事件發生的次數。若再配合 EFDIA 中的時鐘，則同時可得到相對應的事件發生率。下列項目可由 EFDIA 中得到：

(a) 第 i 個位置的失效頻率：$M(P^F_i)/t$

(b) 第 i 個位置的誤差頻率：$M(P^L_i)/t$

(c) 第 i 個位置的維護頻率：$M(P^M_i)/t$

(d) 第 i 個位置的警告頻率：$M(P^W_i)/t$

經由這些發生率可得到下列兩個好處：

(a) 如果第 i 個位置在被偵測到有誤差時即施以維護，則該位置的失效率可被降至最低，從而系統的可靠度則得以提昇。

(b) 所有的發生率均可被記錄成歷史資料，以作為系統失效之統計預測時的依據(如失效頻率、誤差頻率)；維修每個次系統所需的時間亦可獲得(經由維護頻率及警告頻率)。

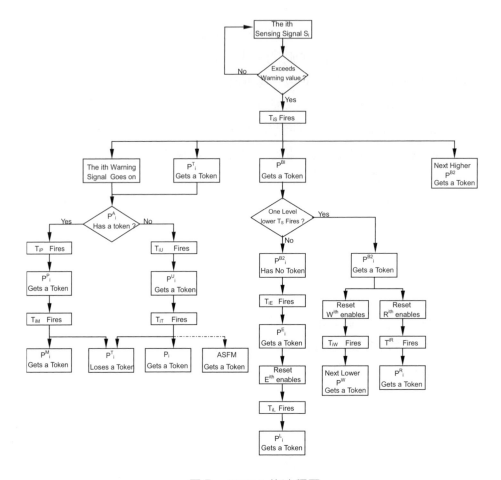

圖八　EFDIA 的流程圖

(5) 系統狀態描述：經由 EFDIA 中每個位置的指示，系統的狀態可清楚地被看見。下列參數即為描述系統狀態之用：

(a) M_k：裴氏網路在第 k 狀態時的標識，$M_k = [\mathrm{M}(P_1), \mathrm{M}(P_2), \ldots, \mathrm{M}(P_n)]^T$

(b) S_k：在第 k 狀態時的感測信號矩陣，$S_k = [\mathrm{M}(S_1), \mathrm{M}(S_2), \ldots, M(S_n)]^T$

(c) L_k：在第 k 狀態時的維護記錄矩陣，$L_k = [\mathrm{M}(PM_1), \mathrm{M}(PM_2), \ldots, \mathrm{M}(PM_n)]^T$

(d) F_k：在第 k 狀態時的失效記錄矩陣，$F_k = [M(P^F_1), M(P^F_2), ..., M(P^F_n)]^T$

(e) E_k：在第 k 狀態時的誤差指示矩陣，$E_k = [M(P^E_1), M(P^E_2), ..., M(P^E_n)]^T$；該矩陣之第 i 項元素值若為 1，即指示該誤差發生在第 i 個位置。

(6) 自動關機或調節：自動關機或調節的功能可經由觸發 ASFM 來提供。

(7) 時間記錄：經由 EFDIA 中的時鐘可記錄每一事件發生的時間，這是進行失效分析時所不可或缺的。

2. 不變量

根據 EFDIA 的性質可推導得到下列不變量：

(1) $M(P^L_i)_k + M(P^R_i)_k = M(P^W_i)_k$
第 i 個位置在第 k 狀態時的警告次數等於誤差出現次數與重置次數的和。

(2) $M(P^L_i)_k + M(P^R_i)_k = M(P^W_i)_k$
第 i 個位置在第 k 狀態時，失效次數加上維護次數等於警告次數。

(3) 在最底層的基本位置處 $M(P^R)_k = 0$ 且 $M(P^L)_k = M(P^W)_k$
因為沒有位置低於基本位置，所以在基本位置處的誤差次數即等於警告次數，且重置次數恆等於零。

(4) $M(P^P_i)_k + M(P^U_i)_k = 1$
因為誤差位置的狀況若不為已維護則為未維護，兩者間彼此為互斥，故第 i 個位置在第 k 狀態時 P^P_i 與 P^U_i 兩者的標識中僅其一為 1，另一為零。

結果與討論

　　將火力發電系統配置 EFDIA 之後的失效分析裴氏網路如圖九所示。這是將圖六中的每一個位置均附加一套 EFDIA 所得之結果，但圖六中所有位置間原有的邏輯關係均仍保留不變。因爲基本位置(也就是 P_1、P_4、及 P_5)位於最底層，所以不須具有測試誤差是否肇因於下一層位置的功能。茲以下述兩個在火力發電系統中的狀況來解說 EFDIA 如何運作：

1. 假設燃油流量的監測信號(即圖九中的 S_1)達到了警告值，則 T_{1S} 發射使得 1st WARNING SIGNAL 作動，並且 P^T_1、P^{B2}_2、及 P^E_1 均分別得到一個標記。$M(P^T_1) = 1$ 代表燃油流量處於誤差狀態，而此情形爲燃油流量介於正常與故障間之過渡狀態。此時距 P_1 失效眞正發生仍有一領先時間。如果在此領先時間內實施了預防維護，則 P^A_1 會產生一個標記使得 T_{1P} 發射，從而將 P^A_1 與 1st WARNING SIGNAL 內之標記一起移至 P^P_1。這表示此時 P_1 次系統正在維護中且 1st WARNING SIGNAL 作動停止。T_{1M} 在預防維護工作結束時發射，使得在 P^P_1 與 P^T_1 中的標記一起移至 P^M_1，表示此誤差狀況已被矯正。P^M_1 的標識(亦即 P_1 的維護次數記錄)會增加 1。另一方面，如果預防維護未能及時實施，則 T_{1U} 發射使得 P^U_1 得到一個標記，所以在 P^U_1 與 P^T_1 中的標記會一起移至 P_1。表示 P_1 失效已發生。同時，P^F_1 會獲得一個標記，也就是說 P_1 的失效次數記錄增加 1。因爲 P_1 與 P_2 間存在原有的邏輯關係，P_1 失效將導致 P_2 的監測信號(即 S_2)達到警告值。所以 2nd WARNING SIGNAL 會作動且 P^T_2、P^{B2}_3、及 P^{B1}_2 均分別得到一個標記。此時 T_{2E} 因 P^{B2}_2 中的標記而被禁止發射。所以在 P^{B1}_2 及 P^{B2}_2 中的標記將在分別觸發 2nd RESET R 及 2nd RESET W 後，分別移至 P^R_2 及 P^W_1。故此誤差狀況被定位在 P_1，$M(P^L_2)$ 並不會增加。

2. 假設連軸轉速的監測信號(即 S_7)達到了警告值，但同時 S_3、S_4、及 S_6 均處於正常狀態。此時 T_{7S} 發射使得 7th WARNING SIGNAL 作動，並且 P^T_7、P^{B1}_7、以及 P^{B2}_8 均分別得到一個標記。接著將與前述 S_1 達到警告值的情形類似，如果 P_7 的預防維護及時實施則 $M(P^M_7)$ 增加 1，否則 P_7 失效會發生

且使得 $M(P^F_7)$ 增加 1。然而，P^{B2}_7、P^{B3}_7、及 P^{B4}_7 均爲空位置，其內均沒有標記，所以 T_{7E} 可發射使得 P^E_7 得到一個標記。其結果是在觸動 7th RESET E 後 P_7 的誤差次數記錄，即 $M(P^L_7)$ 增加 1。

ASFM 是藉由自動關機或自動調節機構來防止更高層的故障或系統當機，ASFM 應配置在裴氏網路中會造成安全問題的位置處。在本例子中由 P_1 到 P_8 均安排有 ASFM。

結　論

本文以火力發電系統爲例子，利用混合式裴氏網路建模法，另配合參數趨勢及故障樹分析，提出了一以預防保養爲目的之早期失效偵測與隔離的方法。使用本方法首先必須建立系統之失效分析裴氏網路，其次是取得失效分析裴氏網路中每一位置的趨勢圖，以便設定該位置之門檻值與警告值。若能滿足此二前提，則本法可適用於任何系統。本文所提出之裴氏網路法不但可達到故障診斷時的早期失效偵測與隔離的要求，而且具有事件計數、系統狀態描述、以及自動關機或自動調節等功能。這些功能對於系統狀態正常與否之監視及預防維護策略之安排均非常有用。

參考文獻

Baccigalupi, A., Bernieri, A. and Pietrosanto, A., "A digital-signal-processor-based measurement system for on-line fault detection", *IEEE Transactions on Instrumentation and Measurement*, Vol. 46, No. 3, pp. 731-736 (1997).

Barna, G. G., "Automatic problem detection and documentation for a plasma etch reactor", *IEEE Transactions on Semiconductor Manufacturing*, Vol. 5, No. 1, pp. 56-59 (1992).

David, R. and Alla, H., "Petri nets for modeling of dynamic systems-A survey", *Automatica*, Vol. 30, No. 2, pp. 175-202 (1994).

Hura, G. S. and Atwood, J. W., "The use of Petri nets to analyze coherent fault trees", *IEEE Transactions on Reliability*, Vol. 37, No. 5, pp. 469-474 (1988).

IEC 50(191), *International Electrotechnical Vocabulary (IEV)*, Chapter 191—Dependability and quality of service, International Electrotechnical Commission, Geneva (1990).

Patton, J. D., Jr., *Preventive Maintenance*, Instrument Society of America (1983).

Rao, S. S., *Reliability-Based Design*, McGraw-Hill, New York (1992).

Rausand, M. and Oien, K., "The basic concept of failure analysis", *Reliability Engineering and System Safety*, Vol. 53, pp. 73-83 (1996).

Schneeweiss, W. G., "Mean time to first failure of repairable systems with one cold spare", *IEEE Transactions on Reliability*, Vol. 44, No. 4, pp. 567-574 (1995).

Yang, S. K. and Liu, T.S., "A Petri net approach to early failure detection and isolation for preventive maintenance", *Quality and Reliability Engineering International*, Vol. 14, pp. 1-12 (1998).

蔡明三 譯，"由維護性談機械可靠性設計"，機械月刊，第八卷第六期，pp. 145-149 民國 71 年六月

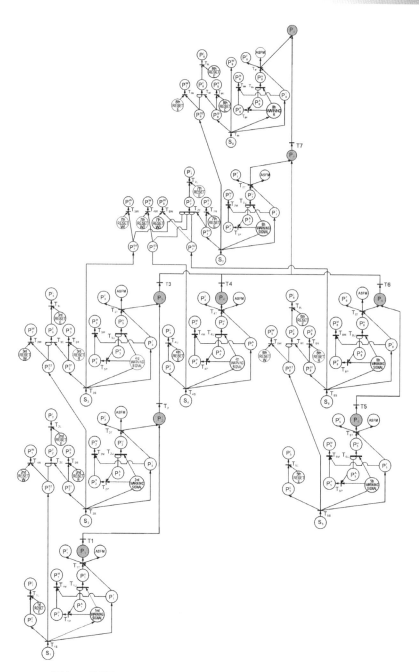

圖九　配置 EFDIA 的火力發電系統之失效分析裴氏網路

附錄 D

IMPLEMENTATION OF PETRI NETS USING A FIELD PROGRAMMABLE GATE ARRAY

SUMMARY

Although Petri nets have various capabilities, the Petri net approach is done on papers. A Field Programmable Gate Array (FPGA) is implemented in this study, so as to realize basic Petri net symbols, logic structures in Petri nets, and specific functions for Petri nets by logic circuits. To exemplify, a Petri net for an Early Failure Detection and Isolation Arrangement (EFDIA) is implemented to an ASIC on a Xilinx Demonstration Board as an example. This ASIC is verified by three simulations dealing with three different failure scenarios of a system, and the ASIC functions identically as that the EFDIA Petri net performs. Accordingly, not only the EFDIA Petri net but also any specific function Petri nets can be implemented by FPGA circuits.

＊**KEY WORDS** ： Petri nets implementation; FPGA; preventive maintenance; logic circuit; ASIC

1. INTRODUCTION

Sequential machines are composed of sequential logic circuits. A sequential machine operates according to a set of sequential conditions. These conditions constitute states of the machine. Therefore, sequential machines are called state machines [1]. A state machine is a synchronous machine if transitions between states are driven by clock pulses. On the other hand, asynchronous machines are driven by changes of inputs, i.e. transitions between states may occur at any time. There are two categories of state machines, namely Moore machines and Mealy machines. The output of a Moore machine is only related to states of the machine, whereas a Mealy machine is related to states and inputs of the machine [2]. State machines have been widely used in design work for many fields due to systematic hardware design methods and tangible implementation models. They support not only synchronous but also asynchronous implementations, which are efficient and easy to design [3].

Petri nets [4] belong to a state machine. Petri nets offer good modeling capability in parallelism and synchronization. Accordingly, they are suitable to perform modeling, analysis, verification, reduction, synthesis [3], etc. A system can be modeled into Petri nets to express not only static behaviors, such as logical relations between components of the system, but also dynamic behaviors such as operating sequence or failure occurrence of the system. Although Petri nets have various capabilities, the Petri net approach is done on papers after all. Now that Petri nets are state machines, it is feasible to realize Petri nets to perform those capabilities. Petri nets can be represented by software or hardware approaches. Software implementation for Petri nets using computer language, e.g. Prolog and CSPL, usually takes long time [5]. Hardware implementation is to realize state machines that are converted from Petri nets to logic circuits. Hardware implementation can be done by choosing one adequate device in the logic device families [6] according to the application, the complexity, and properties of the Petri net.

The advent of the semiconductor manufacturing technology, enables higher density integrated circuits (ICs) such as VLSI (Very Large Scale IC) or ULSI (Ultra Large Scale IC) to appear. Nowadays, ICs become smaller and more powerful but faster and cheaper. As a result, application specific integrated circuits (ASICs) are widely used. In practice, Petri nets can be implemented as ASICs, so as to perform specific functions without user intervention. Early failure detection and isolation depicted in [7], for example, can be implemented as an ASIC. Furthermore, the ASIC with early failure detection and isolation function can be equipped on a system to assist the decision making of preventive maintenance (PM) for the system. Generally speaking, there are four different approaches to ASIC design [8]:

1. Full Custom. All design works, from logic design to lay out and wire routing, are done by a designer according to requests from the customer. This method has the widest design flexibility and the highest efficiency of the wafer. On the contrary, it needs the longest design time, well-experienced personnel, and the highest non-recurring engineering (NRE) cost. Moreover, the product function is specific and with low interchangeability.

2. Standard Cell. Designers perform design works by using circuits in the existing cell library. If the needed circuit is not available, it will be designed or purchased and then added to the cell library. In contrast to full custom design, this method saves design time and increases design flexibility.

3. Gate Array. Using the semi-manufactured wafer, which has logic gates in array form, to execute design works merely by designing the metal layer so as to connect the gates to the desired IC. This method reduces the use of masks for many layers of the wafer such that the NRE cost is lowered. But the wafer area is increased due to the fixed arrangement of the array.

4. Programmable Logic Device (PLD). PLDs are manufactured of existing structures such as RAM, ROM, or PLA [6], which enables designers to write (for EEPROM based components) or download (for SRAM based components) the designed circuits to the PLD so as to perform on-line verification. Consequently, it is not necessary to manufacture a prototype IC by wafer factory and package factory to verify its function. Hence, this method saves the most design time and the NRE cost. FPGAs and CPLDs are main tools for this type of design.

2. FIELD PROGRAMMABLE GATE ARRAY

Mainly because of the programmable capability, Field Programmable Gate Array (FPGA) is becoming popular among not only industry but also academics. Main features of FPGAs [6] are stated below.

1. Field programmable. FPGAs can be programmed by end users. This feature saves the time spent in factories including manufacturing and waiting.

2. Reprogrammable. This feature makes FPGAs to be suitable for teaching and research.

3. Rapid prototyping. The downloaded FGPA itself is a prototype of the newly developing product.

4. No IC-test and NRE cost. FPGAs avoid the NRE cost and promote the efficiency of debugging by vendor supported simulation software or simulator.

5. Fast time to market. It puts newly developed merchandise on market in short time so as to catch up market requirements.

6. In-circuit design verification.

Based on the above features, FPGA is suitable for hardware implementation of Petri nets. This study employs Xilinx FPGA [9] as the design tool to implement Petri nets.

3. CIRCUITS CONVERTED FROM PETRI NETS

By using Xilinx Foundation [9], each of the Petri net converted circuits can be generated as a macro symbol for the schematic toolbox. Detail circuits of corresponding macro symbols can be observed by hierarchy push-and-pop functions.

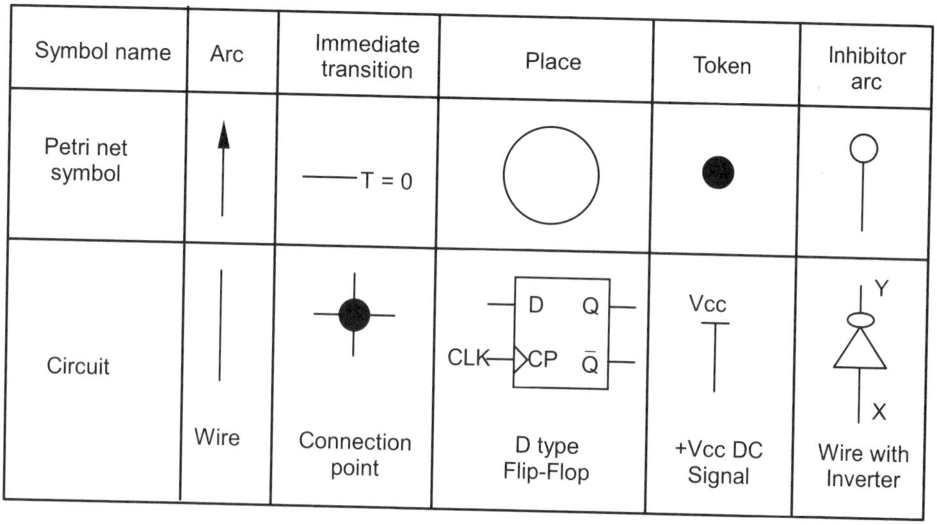

Symbol name	Arc	Immediate transition	Place	Token	Inhibitor arc
Petri net symbol	↑	——T = 0	◯	●	⊖
Circuit	Wire	Connection point	D type Flip-Flop	+Vcc DC Signal	Wire with Inverter

Figure 1. Corresponding circuits for basic Petri net symbols

3.1 Petri Net Symbols

The following five items are basic symbols for Petri nets [10]. Corresponding circuits are listed in Figure 1.

1. *Place*, drawn as a circle, denotes event. A place can be converted to a D type flip-flop, which represents the associated event occurrence by output high Q. Q is high if D is high at the rising edge of the clock pulses.

2. *Token*, drawn as a dot, contained in places, denoting the data. In logic circuits, token can be represented as a logic high signal.

3. *Arc*, drawn as an arrow, between places and transitions. Arcs are connection wires between components.

4. *Immediate transition*, drawn as a thin bar, denoting event transfer with no delay time. Therefore, a connection point represents it.

5. *Inhibitor arc*, drawn as a line with a circle end, between places and transitions. An inhibitor arc can be converted to a connection wire with an inverter. It inverts the relation between input (X) and output (Y).

3.2 Structures of Basic Logic Relations

Basic logic structures of Petri nets and associated logic circuits are presented below, which are illustrated in Figure 2 where X, Y, and CLK represents input place, output place and clock, respectively. All D type flip-flops are positive edge triggered.

1. TRANSFER. The token (logic high signal in the circuit) in X transfers to Y through transition T without delay time.
2. AND. Both X1 and X2 are signal high then Y is high.
3. OR. Either X1 or X2 is high then Y is high.

Logic relation	TRANSFER	AND	OR
Description	If X then Y	If X1 and X2 then Y	If X1 or X2 then Y
Boolean function	Y=X	Y=X1*X2	Y=X1+X2
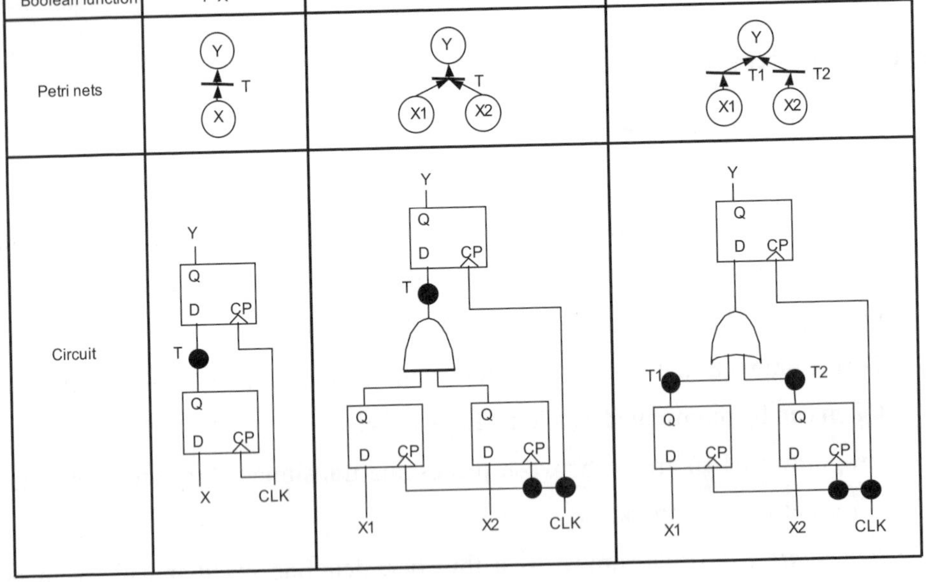 Petri nets			
Circuit			

Figure 2. Structures of basic logic relations for Petrinets and corresponding circuits (a)

4. TRANSFER AND. In this configuration, X feeds high signal to both Y1 and Y2. It describes the situation that X holds, then Y1 and Y2 hold at the same time.

5. TRANSFER OR. X flip-flop triggers either Y1 or Y2 depending on the earlier occurred event between X1 and X2. It is a conditional configuration that was described in [7].

6. IDENTITY. It is a self-content circuit, which supplies a token at anytime for Petri nets.

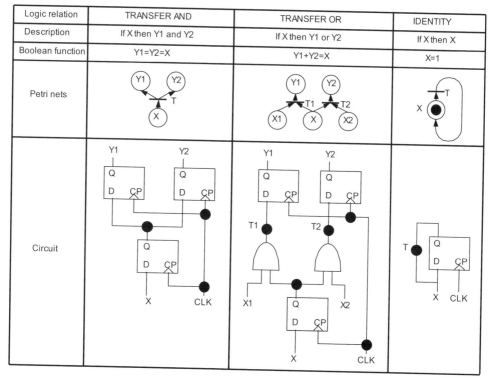

Logic relation	TRANSFER AND	TRANSFER OR	IDENTITY
Description	If X then Y1 and Y2	If X then Y1 or Y2	If X then X
Boolean function	Y1=Y2=X	Y1+Y2=X	X=1
Petri nets			
Circuit			

Figure 2.   Structures of basic logic relations for Petrinets and corresponding circuits (b)

7. INVERT (COMPLEMENT). The relation between X and Y is always inverted.

8. INHIBITION. The condition for high Y is low X1 but high X2. It inhibits Y to be high whenever X1 event occurs.

9. IMPLECATION. Y is triggered either by low X1 or by high X2.

Logic relation	INVERT(COMPLEMENT)	INHIBITION	IMPLECATION
Description	if $\bar{X}$ than Y	if $\bar{X1}$ and X2 then Y	if $\bar{X1}$ or $\bar{X2}$ then Y
Boolean function	$Y = \bar{X}$	$Y = \bar{X1} \times X2$	$Y = \bar{X1} + X2$
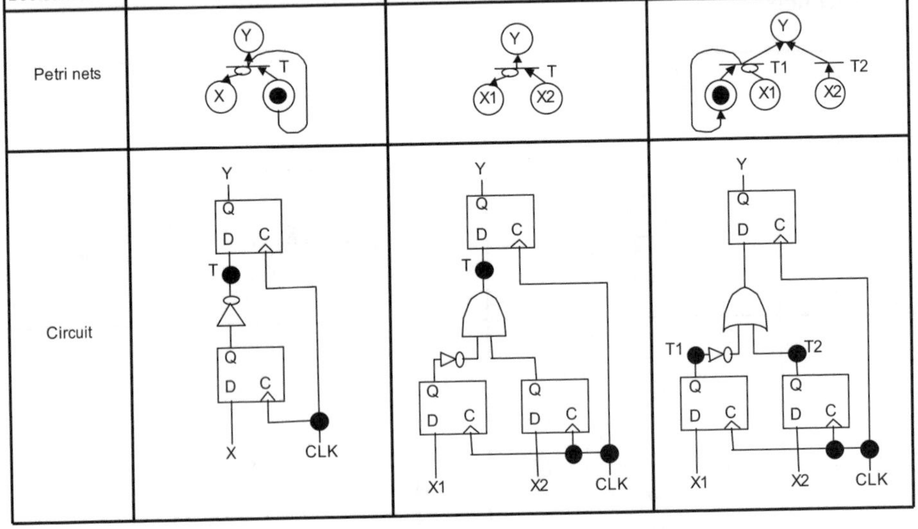 Petri nets			
Circuit			

Figure 2. Structures of basic logic relations for Petrinets and corresponding circuits (c)

10. NAND. Y is high if both X1 and X2 are low. The transition T1 in Petri nets is involved in the NAND gate such that T1 can not be seen in the logic circuit.

11. NOR. Y is triggered when both X1 and X2 are low.

Logic relation	NAND	NOR
Description	If $\overline{X1}$ or $\overline{X2}$ then Y	If $\overline{X1}$ and $\overline{X2}$ then Y
Boolean function	$Y=\overline{X1*X2}$	$Y=\overline{X1+X2}$
Petri nets	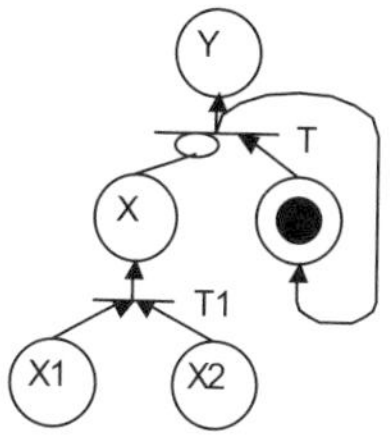	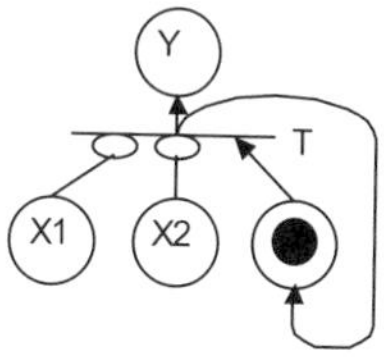
Circuit	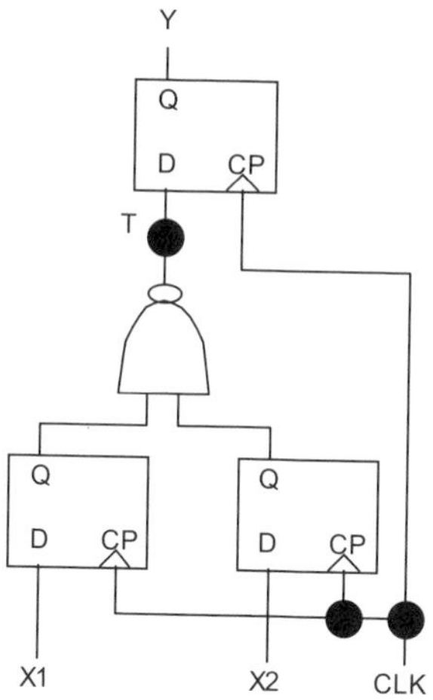	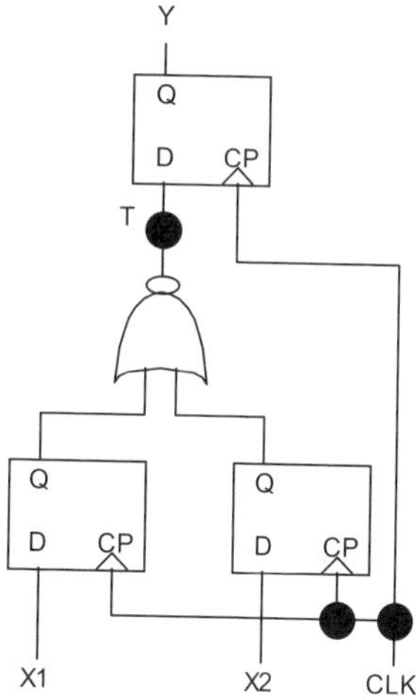

Figure 2. Structures of basic logic relations for Petrinets and corresponding circuits (d)

12. XOR. Y is high if X1 not equals to X2. This configuration is used to verify whether X1 and X2 are not the same.

13. XNOR. XNOR is the complement of XOR, which is used to verify whether X1 and X2 are the same.

Generally, relations between places in a Petri net are constructed by the above logic structures. Each circuit for the structures can be generated into the toolbox as a macro symbol for the ease of Petri net circuit design.

Logic relation	XOR	XNOR
Description	If X1 not equals to X2 then Y	If X1 equals to X2 then Y
Boolean function	Y=X1*$\overline{X2}$+$\overline{X1}$*X2	Y=X1*X2+$\overline{X1}$*$\overline{X2}$
Petri nets		
Circuit		

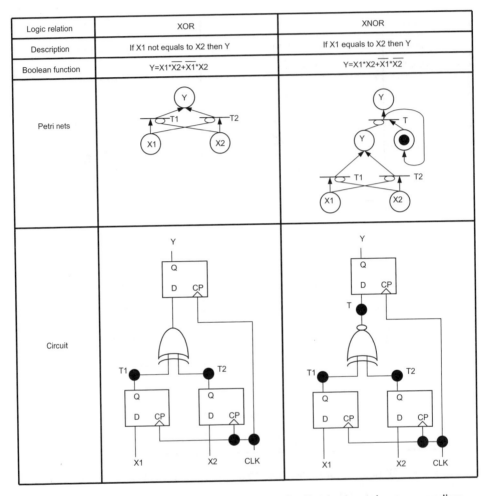

Figure 2.   Structures of basic logic relations for Petrinets and corresponding circuits (e)

3.3   Specific Function Arrangements

1.   *Reset.* This function is used to release the token that is hold in a place by generating a token to fire the output transition of the place. Since a token is implemented by a logic high signal, the reset function can be implemented as a push button with a Vcc input.

2. *Counter*. It is used to count and record event occurrence times. There are various types of counters in Xilinx XACT libraries [11]. In a Petri net dealing with system failures, the counter should count up to a sufficient number and be able to be cleared asynchronously. Therefore, a 4-bit cascade binary counter with clock enable and asynchronous clear (CB4CE) is adopted in this study. The CB4CE is shown in Figure 3 and pin functions are described as follows:

   (1) CE is the clock enable input, which is used to enable the counter itself.

   (2) C stands for the clock.

   (3) Q0, Q1, Q2, and Q3 constitute four data output bits. They increment when the CE is high during the low-to-high clock transition.

   (4) CEO is the counter enable output, which is used to enable the next stage counter.

   (5) TC denotes terminal count. It is high when all Qs are high.

   (6) CLR is the asynchronous clear. When CLR is high, all other outputs are ignored and all Qs and TC outputs go to logic level zero, independent of clock transition.

3. *Timed Transition*. It denotes event transfer with delay time t. As shown in Figure 4, it is implemented by a timer with delay time t and start-reset functions. The timer output becomes high at t time later than the arrival of a logic high signal at the timer input. To achieve this function, a two-level hierarchy configuration circuit is used. The lower level is a frequency divider, namely FREQDIV15 in this study, dividing the input clock frequency by 15. The FREQDIV15 circuit is shown in Figure 5, where the X74_160 is a 4-bit BCD counter [11]. The FREQDIV15 sends a clock pulse out for every 15 input clock pulses and clears X74_160 at the positive edge of the 16th clock pulse. The FREQDIV15 is generated to a macro symbol, as shown in Figure 6, for the design toolbox of this project file. The upper level circuit of the timer

configuration is shown in Figure 6. There is an existing oscillator in Xilinx XACT library, namely OSC4, which supplies five different frequencies of clock, i.e. 15 Hz, 490 Hz, 16k Hz, 500k Hz, and 8M Hz. The FREQDIV15 outputs a 1 Hz clock by feeding the OSC4 15 Hz clock into the FREQDIV15. The 1 Hz clock is used as a base time to generate the N-sec time delay for timed transition merely follows the FREQDIV15 a MOD-N frequency divider. A MOD-20 frequency divider circuit, for example, follows the FREQDIV15 in Figure 6. The D flip-flop and the AND gate construct a switch to start and stop counting delay time of timed transitions by the IN2 trigger signal and the STOP signal, respectively. The STOP signal also resets the timer. Using the technique similar to the DELAY20 and the five clock frequencies provided by OSC4, a variety of delay times can be implemented.

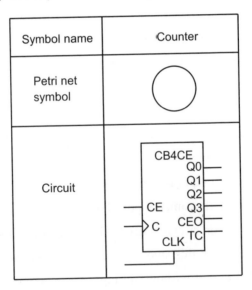

Symbol name	Counter
Petri net symbol	
Circuit	

Figure 3.   Counter circuit converted from Petri net

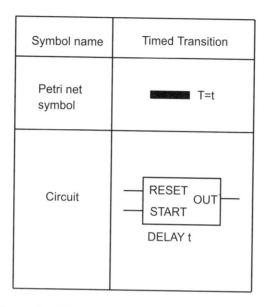

Figure 4. Timer circuit converted from Petri net

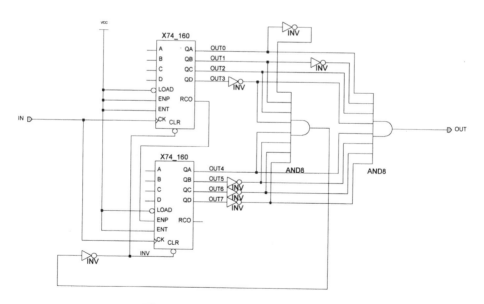

Figure 5. Circuit of FREQDIV15

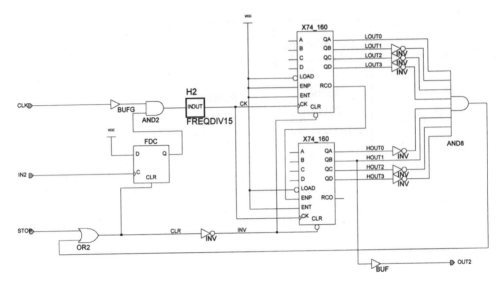

Figure 6.   Circuit of DELAY20

4.   EXAMPLE

The early failure detection and isolation arrangement (EFDIA), depicted in [7] and shown in Figure 7, is employed in this Section to exemplify realization of Petri nets using FPGA. All necessary components have been constructed in Section 3.

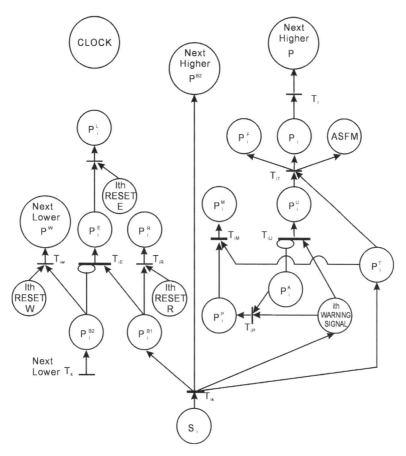

Figure 7.   Early Failure Detection and Isolation Arrangement(EFDIA)

4.1   Circuit of EFDIA

Using circuits described in Section 3, the logic circuit of the EFDIA is constructed as shown in Figure 8. Each of H3 and H4 in Figure 8 is a timer composed of DELAY20. The EFDIA circuit can be integrated into a 39-pins ASIC. Figure 9 shows the macro symbol for EFDIA. Hence, the EFDIA Petri net is realized to become an ASIC as long as downloading this EFDIA macro to a Xilinx FPGA board.

The correspondence between EFDIA pin names (Figure 9) and EFDIA Petri net symbol names (Figure 7) are listed below.

1. Input pins

    (1) CPI-1W: clear signal, which is implicit in Figure 7, for Next Lower P^W counter

    (2) TI-1S: Next Lower T_S

    (3) SIN: S_i

    (4) PIA: P^A_i

    (5) IRW: ith Reset W

    (6) IRR: ith Reset R

    (7) IRE: ith Reset E

    (8) CPIR: clear signal, which is implicit in Figure 7, for P^R_i counter

    (9) CPIM: clear signal, which is implicit in Figure 7, for P^M_i counter

    (10) CPIL: clear signal, which is implicit in Figure 7, for P^L_i counter

    (11) CPIF: clear signal, which is implicit in Figure 7, for P^F_i counter

2. Output pins

    (1) PIT: P^T_i

    (2) PIB1: P^{B1}_i

    (3) IWS: ith WARNING SIGNAL

    (4) PIE: P^E_i

    (5) PI-1WQ0~PI-1WQ3: Next Lower P^W counter

    (6) PIRQ0~PIRQ3: P^R_i counter

    (7) PILQ0~PILQ3: P^L_i counter

    (8) PIFQ0~PIFQ3: P^F_i counter

    (9) PIMQ0~PIMQ3: P^M_i counter

    (10) PI: P_i

    (11) PIB2: P^{B2}_i

    (12) NHPB2: Next Higher P^{B2}

(13) ASFM: ASFM

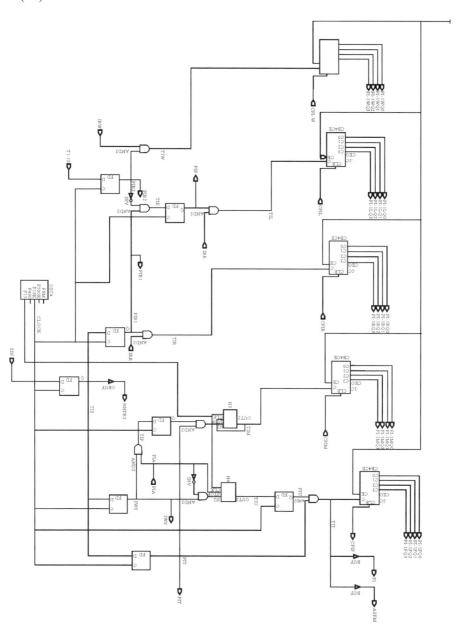

Figure 8.　Circuit of EFDIA

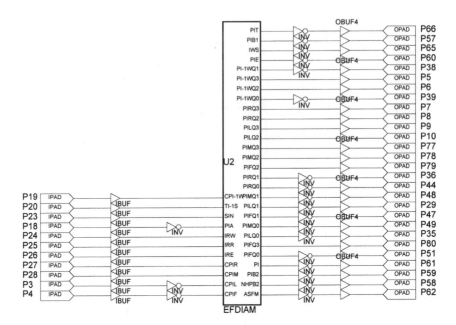

Figure 9.   EFDIA macro symbol

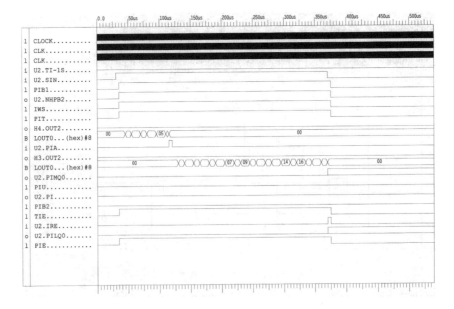

Figure 10.   Timing diagram for TI-1S is low and PIA is high simulation

4.2   Simulations for EFDIA

To verify the EFDIA logic circuit, i.e. Figure 8, the following three failure scenarios are simulated. These simulations are all performed on a Xilinx Logic Simulator.

1.   The TI-1S is low and the PIA is high.

This simulation accounts for that the error signal is not caused by the next lower subsystem (module) and PM action for the malfunctioned subsystem (module) takes place in time. The resultant timing diagram for this simulation is shown in Figure 10, which is generated by a Xilinx Wave Viewer. A logic high signal is sent to the SIN by the simulator, which representing the monitored signal of the ith place exceeds the prescribed warning value. Subsequently, each of PIB1, NHPB2, IWS, and PIT is triggered. The H4 timer, its output is the H4OUT2 as shown in Figure 6, starts counting. The delay time of the H4 timer, which is denoted by T_{iU} in Figure 7, represents the time between the warning value and the maximum allowed value for a system performance, i.e. the maintenance lead time [7]. If the PM action is taken before the system failure occurs, i.e. before the H4 timer completes counting, the PIA sends a signal to start the H3 timer and to stop the H4 timer at the same time. The delay time of the H3 timer represents the time for the PM action to complete. As shown in Figure 10, for example, the PIA signal arises at the 7th count of the H4 timer. Due to the inhibition configuration of T_{iU} transition, which is constructed by the ith WARNING SIGNAL and the P^A_i, TIU will not be triggered resulting from the H4 timer stops counting once the PIA is high. Hence, PI remains low form the beginning to the end of this case, i.e. the system failure is avoided. The output signal of the H3 timer is the H3OUT2 in Figure 10. After H3 counts 20, i.e. after the correcting work for the system error is completed, the PIM counter is triggered such that the PIMQ0 becomes one. At the same time, the SIN becomes low and each of PIB1, NHPB2, IWS, and PIT in turn becomes low. On the other side of the EFDIA, since PIB2 is low and PIB1 is high, TIE becomes high that triggers PIL counter such that the PILQ0 becomes one when IRE generates a trigger signal. Consequently, the error times log number of the ith subsystem increases by one. Since PIE is the error indication flag for the EFDIA, PIE high indicating the error signal is arisen from the ith subsystem but not the next lower one. The PIE timing curve is shown at the bottom in Figure 10.

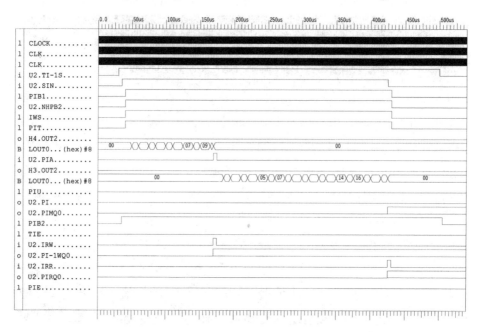

Figure 11.   Timing diagram for TI-1S is high and PIA is high simulation

2.   The TI-1S is high and the PIA is high.

This simulation deals with an error signal arising from the next lower subsystem but not the ith subsystem itself. The timing diagram for this simulation is shown in Figure 11. Due to logic relations constrained by the Petri net dealing with system failures [12], the SIN becomes high after the high TI-1S signal. The high SIN signal triggers each of PIB1, NHPB2, IWS, and PIT to become high. The H4 timer is triggered to count by low PIA and high IWS. The low PIE indicates that the error is not located in the ith subsystem itself. Triggering the IRW enables the PI-1WQ0 to become one. Hence, the warning times log number of the next lower subsystem increases by one. Since the PIE is low, the PIRQ0 can be triggered by an IRR signal and high PIB1. In this simulation, the PIA signal triggers the H3 timer to count after an inspection for the ith subsystem but not a maintenance action. Once the H3 timer starts counting, the H4 timer counting stops. When the PIMQ0 becomes one, i.e. the maintenance log number of the ith subsystem increases by one, the SIN resumes low from high. The TI-1S resets to low after the correcting work for the next lower subsystem is completed. The PI signal remains low in this simulation, i.e. a P_i failure never occurs.

3. The TI-1 is low and the PIA is low.

This last simulation accounts for error existing in the ith subsystem but without PM. The PIE is triggered to high due to high PIB1 and low PIB2. High PIE implies that the error is located in the ith subsystem. The H4 timer is triggered to count by the warning signal IWS and low PIA. Since the PM action is not taken during the maintenance lead-time, i.e. the delay time of the H4 timer, the PIU becomes high after the H4 timer counts 20. As a result, the PI becomes high, representing P_i failure occurs. The failure times log number of this subsystem increases by one, i.e. the PIFQ0 becomes one at this moment. The PILQ0 is also triggered by the IRE signal to increase the error times log number by one for the ith subsystem. The timing diagram for this simulation is shown in Figure 12.

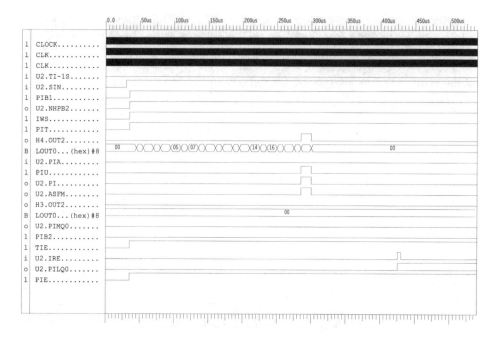

Figure 12.   Timing diagram for TI-1S is low and PIA is low simulation

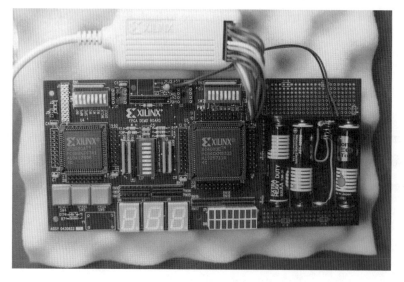

Figure 13.   The downloaded Demonstration Board

Figure 13.　The downloaded Demonstration Board(續)

4.3　Implementation

The EFDIA logic circuit is implemented by downloading its schematic diagram to a Xilinx FPGA Demonstration Board. The board is a stand-alone board for experimenting and developing prototypes using the Xilinx FPGA architecture. Two FPGA devices, namely XC3020A and XC4003E have been installed on the board. The XC4003E has higher density and more input/output blocks and flip-flops than the XC3020A has. [6] Hence, the XC4003E is adopted in this study to implement EFDIA. The configuration of the XC4003E for implementing the EFDIA is described as follows.

1. *Power supply*. The power for the Demonstration Board is supplied by a battery set, which has 3 AA(UM-3) batteries in series to supply +5 volts through the connector J9 of the board.

2. *Downloading interface.* The EFDIA schematic diagram for configuring the XC4003E is downloaded from a personal computer through an Xchecker cable [13] which connects either the COM1 or COM2 port of the computer to the J2 connector of the board.

3. *Input terminals.* Switches SW3, SW4 and SW5 provide input signals for the XC4003E to implement the EFDIA circuits. The SW3 is a switch set with eight switches connecting to eight general-purpose inputs on XC4003E input pins. An XC4003E input pin is set to logic 1 when the corresponding switch is on, and logic 0 when the corresponding switch is off. The SW4, namely Reset Pushbutton, can apply an active-Low reset signal to the XC4003E via pin 56 when the SW2-7 switch is on. As for the SW5, namely Spare Pushbutton, applies also an active-Low signal to the XC4003E via pin 18.

4. *Output terminals.* Three seven-segment displays are included with the U6 connect to the XC3020A, and U7 and U8 connect to the XC4003E. Each LED segment is turned on by driving the corresponding FPGA pin Low with logic 0. Decimal points serve as state and error indicators. Besides, there are eight LEDs connected to the I/O pins in each FPGA. LEDs D1 through D8 connected to the XC3020A, while D9 through D16 connect to the XC4003E. Each LED is also turned on by driving its corresponding FPGA pin Low with logic 0. There are extra 16 I/O lines that connect each FPGA.

Figure 13 shows two pictures of the downloaded Demonstration Board, and the I/O assignment on the Demonstration Board for the EFDIA implementation is shown in Figure 14.

4.4 Results

According to timing diagrams for the simulations in Section 4.2, i.e. Figures 10 to 12, and operations on the downloaded Demonstration Board, EFDIA logic circuit functions identically to the EFDIA Petri net. All the capabilities of the

EFDIA Petri net including alarm, early failure detection, fault isolation, event count, system state description, and automatic shutdown or regulation are preserved in the 39-pins ASIC depicted in Section 4.1. Hence, the Petri nets have been realized by FPGA circuits.

5. APPLICATIONS

Capabilities of the EFDIA are very useful for health monitoring, on-line failure prognostics, and preventive maintenance of a system. No matter what scale the system is, the EFDIA is applicable. From large systems such as power plants and chemical plants to smaller systems like automobiles and machinery, a set of personal computer, or a CD-ROM, are within the scope of EFDIA application. Taking an automobile as an example, a warning light shown on the panel denotes that a failure with prescribed threshold is going to occur and where the cause comes from, by equipping an EFDIA ASIC to each of sensing points corresponding to places in the Petri net. Consequently, errors in an automobile can be corrected before a failure occurs by preventive maintenance. Accordingly, driving safety can be ensured.

6. CONCLUSIONS

Although Petri nets are suitable to perform modeling, analysis, verification, reduction, synthesis and so on, the Petri net approach is a paper work after all. Since Petri nets are state machines, it is possible to be realized to perform those capabilities. This paper has presented hardware implementation of Petri nets, including basic symbols, basic logic structures, and specific functions for Petri nets. Besides, the Petri net with early failure detection and isolation functions has been implemented to a 39-pins ASIC on a Xilinx Demonstration Board as an example. This ASIC was verified by three simulations dealing with three different failure scenarios of a system. The 39-pins ASIC functions identically as that the EFDIA

Petri net performs. Since Petri nets offer a convenient modeling paradigm and other various functions, not only the EFDIA Petri net but also any specific function Petri nets can be implemented by FPGA circuits.

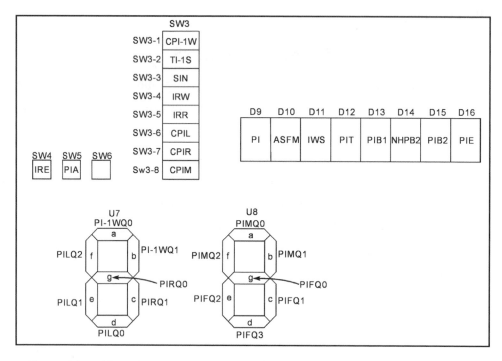

Figure 14.   I/O assignment on Demonstration Board for EFDIA implementation

REFERENCES

1.   A. W. Shaw, *Logic Circuit Design*, Fort worth Saunder College Publishing, Fort Worth, 1993.

2.   C. H. Roth Jr, *Fundamentals of Logic Design*, 4th edition, International Thomson Publishing, Asia, 1995.

3.   N. Chang, W. H. Kwon, and J. Park, 'FPGA-based implementation of synchronous Petri nets', *Proceedings of the IECON'96, International*

Conference on Industrial Electronics, Control, and Instrumentation, 1996, pp. 469-474.

4. J. L. Peterson, *Petri net Theory and the Modeling of Systems*, Prentice-Hall, Englewood Cliffs, New Jersey, 1981.

5. A. D. Stefano and O. Mirabella, 'A fast sequence control device based on enhanced Petri nets', *Microprocessors and Microsystems*, **15**, 179-186 (1991).

6. CIC, *Training Course 12*, Chip Implementation Center of Nation Science Council, Republic of China, 1998.

7. S. K. Yang, and T. S. Liu, 'A Petri net approach to early failure detection and isolation for preventive maintenance', *Quality and Reliability Engineering International*, **14**, 319-330 (1998).

8. J. Schroeter, *Surviving The ASIC Experience*, Prentice-Hall, Englewood Cliffs, New Jersey, 1992.

9. Xilinx, *Foundation series quick start guide version F1.4*, The Programmable Logic Company, San Jose, 1998.

10. R. David and H. Alla, 'Petri nets for modeling of dynamic systems-A survey', *Automatica*, **30**, (2), 175-202 (1994).

11. Xilinx, *XACT Libraries Guide*, The Programmable Logic Company, San Jose, 1994.

12. S. K. Yang, and T. S. Liu, 'Failure analysis for an airbag inflator by Petri nets', *Quality and Reliability Engineering International*, **13**, 139-151 (1997).

13. Xilinx, *Hardware User Guide*, The Programmable Logic Company, San Jose, 1998.

附錄 E

CONDITION-BASED FAILURE PREDICTION

SUMMARY

Failure can be prevented in time by preventive maintenance (PM) so as to promote reliability only if failures can be early predicted. Time-based and condition-based maintenance are two major approaches for PM. No matter which approach is adopted for PM, whether a failure can be early detected or even predicted is the key point. This chapter presents a failure prediction method for PM by state estimation using the Kalman filter on a DC motor. The prediction consists of a simulation on a computer and an experiment on the DC motor. In the simulation, an exponential attenuator is placed at the output end of the motor model to simulate aging failures by monitoring one of the state variables, i.e. rotating speed of the motor. Failure times were generated by Monte Carlo simulation and predicted by the Kalman filter. One-step-ahead and two-step-ahead predictions are conducted. Resultant prediction errors are sufficiently small in both predictions. In the experiment, the rotating speed of the motor was uninterruptedly measured and recorded per 5 minutes for 80 days. The measured data are used to execute Kalman prediction and to verify the prediction accuracy. Resultant prediction errors are acceptable. However, the shorter the increment time for every step in Kalman prediction uses, the higher prediction accuracy it achieves. Consequently, failure can be prevented in time so as to promote reliability by state estimation for predictive maintenance using the Kalman filter.

∗**KEY WORDS** ： Kalman filter; failure prediction; state estimation; preventive maintenance; DC motor

Nomenclature

$(\bullet)k$ = The value of ($\bullet$) at time kT

$(\hat{.})_{a/b}$ = The estimate of (.) at time aT based on all known information about the process up to time bT.

A = A matrix

Ac = Coefficient matrix of the state equation for a continuous system

Ad = Coefficient matrix of the state equation for a discrete system

A^T = Transpose matrix of A

A^{-1} = Inverse matrix of A

B = Damping coefficient

B_c = Coefficient matrix of the state equation for a continuous system

B_d = Coefficient matrix of the state equation for a discrete system

B_k = Coefficient matrix for the input term of a discrete state equation

C = A matrix

C_c = Coefficient matrix of the state equation for a continuous system

C_d = Coefficient matrix of the state equation for a discrete system

D_c = Coefficient matrix of the state equation for a continuous system

D_d = Coefficient matrix of the state equation for a discrete system

E = Applied voltage

E_r = Estimation error

E_k = Matrix giving the ideal (noiseless)connection between the measurement and the state vector

i_a = Armature winding current

J = Moment of inertia of rotor and load

k_b = Back emf constant

K_k = Kalman gain

K_T = Motor torque constant

L_a = Armature winding inductance

L^{-1} = The inverse Laplace transform

$P_{k/k-1}$ = Estimation error covariance matrix

Q_k = Covariance matrices for disturbance

R = Armature winding resistance

R_k = Covariance matrices for noise

t = Time variable

T = Motor output torque

T = Increment time for every step in Kalman prediction

U_k = Control input of a discrete state equation at state k

V = Variation of the estimated rotating speed

V_k = Noise, measurement error vector

It is assumed to be a white sequence with known covariance.

W_k = Disturbance, system stochastic input vector
It is assumed to be a white sequence with known covariance and having zero crosscorrelation with V_k sequence.

x, X = Variable of a distribution function

X_{D0} = Initial states resulting from deterministic input

X_k = System state vector at state k

X_{S0} = Initial states resulting from stochastic input

Y_k = System output vector at state k

Z_k = Output measurement vector

θ = Motor angle displacement

$\dot{\theta}$ = Motor rotating speed

μ = Mean value of a distribution function

σ = Standard deviation of a distribution function

Φ_k = Matrix relating X_k to X_{k+1} in the absence of a forcing function
It is the state transition matrix if X_k is sampled from a continuous process.

1.   INTRODUCTION

High quality and excellent performance of a system are always goals which engineers strive to achieve. Reliability engineering integrates quality and performance from the beginning to the end of a system life [1]. Therefore, reliability can be treated as the time-dimensional quality of a system. Reliability is affected by every stage throughout the system life, including its development, design, production, quality control, shipping, installation, operation, and maintenance. Consequently, paying attention to each of the stages promotes reliability. Specifically, in the onsite operation phase, failures are the main causes of worsened performance and degraded reliability. It is very important for any equipment whose failure may cause severe damage to public safety or financial benefit, such as nuclear power plants, passenger vehicles, or semiconductor production lines. Accordingly, failure avoidance is the main approach to reliability assurance. To achieve failure reduction, an effective maintenance is the best way [1]. There are three main types of maintenance: improvement maintenance (IM), corrective maintenance (CM), and preventive maintenance PM [2]. The purpose of IM is to reduce or eliminate entirely the need for maintenance, i.e. IM is performed at the design phase of a system emphasizing elimination of failures. There are many restrictions for a designer, however, such as space, budget, and market requirements. Usually the reliability of a product is related to its price. On the other hand, CM is the repair performed after failure occurs. PM means all actions intended to keep equipment in good operating condition and to avoid failures [2]. The most common strategy for maintenance is scheduled maintenance, i.e. maintenance is executed by time, by operation times, by material number, or by some other prescribed criteria. Nevertheless, there are at least two drawbacks for this type of maintenance:

1. Criteria on which the scheduled maintenance is based are statistical averages, e.g. mean time to failure (MTTF). It makes the risk unavoidable that a system fails before criteria are exceeded, i.e. a failure may occur unexpectedly.

2. The real duty cycles for certain parts or modules may be longer than those averages, but they are replaced during a scheduled maintenance. It is a waste for the investment.

By contrast, condition monitor is a better and reasonable type of maintenance than scheduled maintenance. However, a failure should be detected prior to its occurrence.

The relationship between error, failure, and fault is illustrated in Fig. 1, and the three terms are defined as follows [3]:

1. An error is a discrepancy between a computed, observed or measured value or condition and the true, specified or theoretically correct value or condition.

2. Failure is an event when a required function is terminated (exceeding the acceptable limits).

3. Fault is the state characterized by inability to perform a required function, excluding the inability during preventive maintenance or other planned actions, or due to lack of external resources.

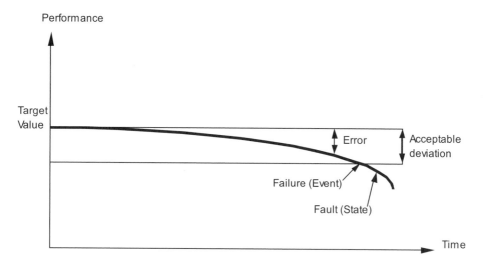

Figure 1.   Error, failure and fault

Based on the above statements, an error is not a failure and a fault is hence a state resulting from a failure. An error is sometimes referred to as an incipient failure [4]. Therefore; PM action is taken when the system is still at an error condition, i.e. within acceptable deviation and before failure occurs. Thus, through the technique of PM, failure can be early detected. Hence, PM is an effective approach to promoting reliability [5]. As aforementioned, time-based and condition-based maintenance are two major approaches for PM. Irrespective of the approach adopted for PM, the key point is whether a failure can be detected early or even predicted.

Many methods have been proposed for failure prediction such as statistical knowledge of the reliability parameters [6, 7], neural network studies [8], and understanding the failure mechanism of damaged products [9]. Fault detection based on modeling and estimation is one of the methods [10]. The Kalman filter is useful not only for state estimation but also for state prediction. It has been widely used in different fields during the past decades, such as on-line failure detection [11], real time prediction of vehicle motion [12], and prediction for maneuvering target trajectories [13]. The Kalman filter is a linear, discrete-time, and finite-dimensional system [14]. Its appearance is a copy of the system that is estimated. Inputs of the filter include the control signal and the difference value between measured and estimated state variables. Actual values of the event acquired by the monitor sensor are fed into the corresponding Kalman filter to execute state estimation. By minimizing mean-square estimation errors, the optimal estimate can be derived. Based upon the current state, the Kalman filter provides a predicted value of the next state for the corresponding event at every time interval T. As a result, the output of the filter becomes optimal estimates of the next step time state variables. Each event has a prescribed failure threshold, and the predicted value is compared with the prescribed failure threshold to judge whether the monitored event has failed after T or is still within the established threshold. Once the estimated value reaches the threshold, the failure is predicted. Accordingly, the current state is a warning state and the PM needs to be performed. If the predicted future state variables indicate a device is going to fail, then the failure can be prevented in time by PM. However, future state variables should be accurately predicted at a reasonably long time ahead of failure occurrence [10, 15].

This chapter proposes the state estimation and prediction for PM using the Kalman filter. A DC motor is employed as the object to perform the condition-based failure prediction. The prediction consists of two parts. The first part is a simulation on a computer, and the second part is an experiment on the DC motor. In the simulation, failure times were generated by Monte Carlo simulation (MCS) and predicted by the Kalman filter. One-step-ahead and two-step-ahead predictions were conducted. Resultant prediction errors are sufficiently small in both predictions. Even so, the failure prediction was simulated on a computer after all. In the second part of the prediction, a DC motor and a data acquisition system are set to implement the simulation. Rotating speed of the motor is chosen as the major state variable to judge whether the motor is going to fail by state estimation using the Kalman filter. The rotating speed of the motor was uninterruptedly measured and recorded every 5 minutes for 80 days. Instead of simulated data the measured data are used to execute Kalman prediction and to verify the prediction accuracy.

In Sect. 2, a discrete system model with deterministic control input, white noise disturbance and noisy output measurement will be constructed first. Equation formulation for state estimation of the Kalman filter then follows. Deterministic inputs are considered in the formulation. Moreover, equations for N-step-ahead prediction are derived. Sect. 3 presents the transfer function, continuous state model, and the discrete state model of a DC motor that is employed as an example in this chapter. Sect. 4 presents the simulation system with prescribed parameters, Monte Carlo simulation and ARMA model used to generate necessary data for failure prediction simulation, and the exponential attenuator used to simulate aging failure mode. Simulation results and discussions are also described in the section. Sect. 5 presents the experiment setup with related parameters, experimental results and discussions. Sect. 6 concludes this work.

2. KALMAN FILTERING

This section introduces related knowledge for the Kalman filtering that is used in this chapter to perform the failure prediction.

2.1 System Model

The block diagram of a discrete system is shown in Fig. 2. The state equations [14] are:

$$X_{k+1} = \Phi_k X_k + B_k U_k + W_k , \tag{1}$$

$$Y_k = H_k X_k , \tag{2}$$

$$Z_k = Y_k + V_k . \tag{3}$$

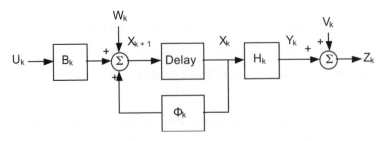

Figure 2. Block diagram of a discrete system

Substituting (2) into (3) yields

$$Z_k = H_k X_k + V_k.$$ (4)

Let $E[X]$ be the expected value of X; thus, covariance matrices for W_k and V_k are given by:

$$E[W_k W_i^T] = \begin{cases} Q_k, & i = k \\ 0, & i \neq k \end{cases},$$ (5)

$$E[V_k V_i^T] = \begin{cases} R_k, & i = k \\ 0, & i \neq k \end{cases},$$ (6)

$$E[W_k V_i^T] = 0, \quad \text{for all } k \text{ and } i.$$ (7)

It follows that both Q_k and R_k are symmetric and positive definite [16].

2.2 State Estimation

State estimation aims to guess the value of X_k by using measured data, i.e. Z_0, Z_1, ... Z_{k-1} Accordingly, $\hat{X}_{k/k-1}$ is called the prior estimate of X, and $\hat{X}_{k/k}$ is called the posterior estimate of X [16]. The prior estimation error is defined as

$$e_{k/k-1} = X_k - \hat{X}_{k/k-1}.$$ (8)

Since W_k and V_k are assumed to be white sequences, the prior estimation error has zero mean. Consequently, the associated error covariance matrix is written as

$$P_{k/k-1} = E[(e_{k/k-1})(e_{k/k-1})^T] = E[(X_k - \hat{X}_{k/k-1})(X_k - \hat{X}_{k/k-1})^T].$$ (9)

The estimation problem begins with no prior measurements. Thus, the stochastic portion of the initial estimate is zero if the stochastic process mean is zero; i.e. $\hat{X}_{0/-1}$ is driven by deterministic input X_{D0} only. It follows from (8) that

$$e_{0/-1} = X_0 - \hat{X}_{0/-1} = X_0 - X_{D0} = X_{S0}. \tag{10}$$

Employing (9) and (10) yields

$$P_{0/-1} = E[X_{S0}X_{S0}^{\ T}]. \tag{11}$$

The Kalman filter is a copy of the original system and is driven by the estimation error and the deterministic input. The block diagram of the filter structure is shown in Fig. 3. The filter is used to improve the prior estimate to be the posterior estimate by the measurement Z_k. A linear blending of the noisy measurement and the prior estimate is written as [16]

$$\hat{X}_{k/k} = \hat{X}_{k/k-1} + K_k(Z_k - H_k\hat{X}_{k/k-1}), \tag{12}$$

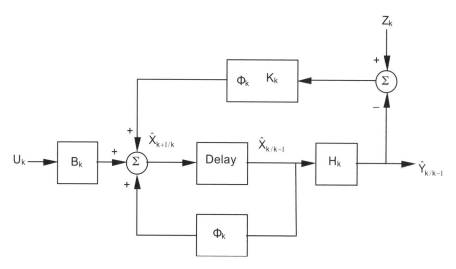

Figure 3.   Block diagram of a Kalman filter

where K_k is a blending factor for this structure. Once the posterior estimate is determined, the posterior estimation error and associated error covariance matrix can be derived as

$$e_{k/k} = X_k - \hat{X}_{k/k},$$ (13)

$$P_{k/k} = E[(e_{k/k})(e_{k/k})^T],$$

$$= E[(X_k - \hat{X}_{k/k})(X_k - \hat{X}_{k/k})^T].$$ (14)

The optimal blending factor is written as [16]

$$K_k = P_{k/k-1} H_k^T (H_k P_{k/k-1} H_k^T + R_k)^{-1}.$$ (15)

This specific K_k, namely, the one that minimizes the mean-square estimation error, is called Kalman gain.

Substituting (15) into (12), the posterior error covariance matrix can be derived as follow:

$$P_{k/k} = P_{k/k-1} - P_{k/k-1} H_k^T (H_k P_{k/k-1} H_k^T + R_k)^{-1} H_k P_{k/k-1}$$

$$= P_{k/k-1} - K_k (H_k P_{k/k-1} H_k^T + R_k) K_k^T$$

$$= (I - K_k H_k) P_{k/k-1}.$$ (16)

As depicted in Fig. 3, the one-step-ahead estimate is formulated as

$$\hat{X}_{k+1/k} = \Phi_k \hat{X}_{k/k-1} + \Phi_k K_k (Z_k - H_k \hat{X}_{k/k-1}) + B_k U_k$$

$$= \Phi_k [\hat{X}_{k/k-1} + K_k (Z_k - H_k \hat{X}_{k/k-1})] + B_k U_k$$

$$= \Phi_k \hat{X}_{k/k} + B_k U_k.$$ (17)

Consequently, the one-step-ahead estimation error is derived as

$$e_{k+1/k} = (\Phi_k X_k + B_k U_k + W_k) - (\Phi_k \hat{X}_{k/k} + B_k U_k)$$

$$= \Phi_k (X_k - \hat{X}_{k/k}) + W_k$$

$$= \Phi_k e_{k/k} + W_k.$$ (18)

In a manner similar to (14), the one-step-ahead error covariance matrix is derived as

$$P_{k+1/k} = E[(\Phi_k e_{k/k} + W_k)(\Phi_k e_{k/k} + W_k)^T]$$
$$= \Phi_k P_{k/k} \Phi_k^T + Q_k. \tag{19}$$

According to the above statements, several remarks for Kalman estimation are concluded as follows.

1. Since K_k is optimal, the posterior estimate $\hat{X}_{k/k}$ is an optimal estimate.

2. Based on (12), (15), (16), (17), and (19), recursive steps for constructing a one-step estimator are summarized in Fig. 4.

3. The recursive loop has two different kinds of updating. Equations (12) and (16) yielding $\hat{X}_{k/k}$ and $P_{k/k}$ from $\hat{X}_{k/k-1}$ and $P_{k/k-1}$ are measurement-update; Equations (17) and (19) projecting $\hat{X}_{k/k}$ and $P_{k/k}$ to $\hat{X}_{k+1/k}$ and $P_{k+1/k}$ are time-update.

4. Initial conditions, i.e. $\hat{X}_{0/-1}$, $P_{0/-1}$, Φ_0, H_0, Q_0, and R_0 have to be known to start recursive steps.

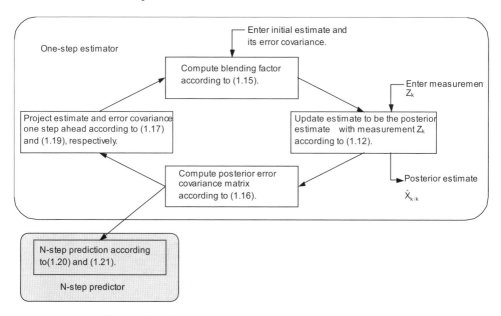

Figure 4.   One-step estimator and N-step predictor

2.3 Prediction

The estimate resulting from recursive steps in Fig. 4 is a one-step-ahead prediction. Based on the posterior estimate, i.e. (12), the state that is N steps ahead of the measurement Z_k can be predicted by using the ARMA (autoregressive and moving average) model [16]. From (17) and (19), equations for N-step-ahead prediction are derived as

$$\hat{X}_{k+N/k} = (\prod_{i=k+N-1}^{k} \Phi_i) \hat{X}_{k/k} + \sum_{m=k}^{k+N-2} [(\prod_{i=k+N-1}^{m+1} \Phi_i) B_m U_m] + B_{k+N-1} U_{k+N-1}, \quad (20)$$

$$P_{k+N/k} = (\prod_{i=k+N-1}^{k} \Phi_i) P_{k/k} (\prod_{j=k}^{k+N-1} \Phi_j^T) + \sum_{m=k}^{k+N-2} [(\prod_{i=k+N-1}^{m+1} \Phi_i) Q_m (\prod_{j=m+1}^{k+N-1} \Phi_j^T)] + Q_{k+N-1}.$$

$$(21)$$

The N-step predictor is an appendage of the one-step estimation loop [16]. It is also shown in Fig. 4. Since the current predicted value is assumed to be the initial value for the next prediction, the more steps the predictor predicts, the larger error it results in.

3. ARMATURE-CONTROLLED DC MOTOR

An armature-controlled DC motor is employed in this section as the physical model to perform failure prediction. The motor circuit representation is shown in Fig. 5.

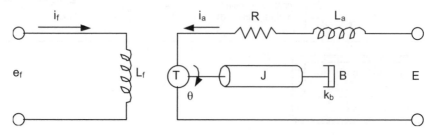

Figure 5.   Circuit representation of DC motor

3.1 Transfer Function

According to properties of a DC motor, the following equations can be formulated [17]:

$$\phi = k_f i_f ,\tag{22}$$

$$T = \frac{ZP}{2\pi a}\phi i_a = k_1\left(k_f i_f\right)i_a = k_T i_a\tag{23}$$

$$e_b = k_b \frac{d\theta}{dt},\tag{24}$$

$$L_a \frac{d}{dt}i_a + Ri_a + e_b = E ,\tag{25}$$

$$J\ddot\theta + B\dot\theta = T ,\tag{26}$$

where $k_1 = \dfrac{ZP}{2\pi a}$ is called the motor constant, and $k_T = k_1\left(k_f i_f\right)$ is the motor torque constant.

Taking the Laplace transform for (24), (25), and (26) results in

$$E_b\left(s\right) = k_b s\theta\left(s\right),\tag{27}$$
$$\left(L_a s + R\right)I_a\left(s\right) = E\left(s\right) - E_b\left(s\right),\tag{28}$$
$$\left(Js^2 + Bs\right)\theta\left(s\right) = T\left(s\right) = k_T I_a\left(s\right).\tag{29}$$

Combining (27), (28), and (29), the transfer function of a DC motor is derived as

$$\frac{\theta\left(s\right)}{E\left(s\right)} = \frac{k_T}{s\left[\left(sL_a + R\right)\left(sJ + B\right) + k_T k_b\right]}.\tag{30}$$

Accordingly, the block diagram of a DC motor can be shown in Fig. 6. If $L_a \approx 0$, (30) can be rewritten as

$$\frac{\theta\left(s\right)}{E\left(s\right)} = \frac{k_m}{s\left(s\tau_m + 1\right)},\tag{31}$$

where $k_m = \dfrac{k_T}{RB + k_T k_b}$ and $\tau_m = \dfrac{RJ}{RB + k_T k_b}$ are called motor gain constant

and motor time constant, respectively.

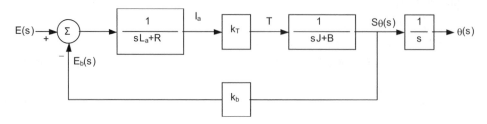

Figure 6.   Block diagram of DC motor

3.2   Continuous State Space Model

Define $\theta,\ \dot{\theta}$, and i_a as state variables, so that the state vector is $X = [\theta\quad \dot{\theta}\quad i_a]^{\mathrm{T}}$. Since

$$\frac{d}{dt}\theta = \dot{\theta}, \tag{32}$$

substituting (23) and (32) into (26) yields

$$\frac{d}{dt}\dot{\theta} = \frac{1}{J}\left(k_T i_a - B\dot{\theta}\right) = \frac{k_T}{J}i_a - \frac{B}{J}\dot{\theta}. \tag{33}$$

Moreover, substituting (24) into (25) yields

$$\frac{d}{dt}i_a = \frac{1}{L_a}\left(E - Ri_a - e_b\right) = \frac{E}{L_a} - \frac{k_b}{L_a}\dot{\theta} - \frac{R}{L_a}i_a. \tag{34}$$

In measurement, the rotating speed $\dot{\theta}$ is the motor output. According to (32), (33) and (34) continuous state equations of the DC motor are

$$\frac{d}{dt}\begin{bmatrix} \theta \\ \dot{\theta} \\ i_a \end{bmatrix} = \begin{bmatrix} 0 & 1 & 0 \\ 0 & -\dfrac{B}{J} & \dfrac{k_T}{J} \\ 0 & -\dfrac{k_b}{L_a} & -\dfrac{R}{L_a} \end{bmatrix}\begin{bmatrix} \theta \\ \dot{\theta} \\ i_a \end{bmatrix} + \begin{bmatrix} 0 \\ 0 \\ \dfrac{1}{L_a} \end{bmatrix} E ,$$

(35)

$$Y = \begin{bmatrix} 0 & 1 & 0 \end{bmatrix}\begin{bmatrix} \theta \\ \dot{\theta} \\ i_a \end{bmatrix} .$$

(36)

3.3 Discrete State Space Model

The general form of state equations for a continuous system reads [18]:

$$\dot{X}(t) = A_c X(t) + B_c U(t)$$
$$Y(t) = C_c X(t) + D_c U(t)$$

(37)

Let $\Phi_c(t) = L^{-1}[(sI - A_c)^{-1}]$ be the state transition matrix for (37). The discrete state equations sampled from (37) by a Sample-and-Hold with time interval T seconds are as follows [19]:

$$X_{k+1} = AX_k + BU_k , \quad Y_k = CX_k + DU_k ,$$

where

$$A = \Phi_c(T) ,$$

(38)

$$B = [\int_0^T \Phi_c(\tau)d\tau]B_c ,$$

(39)

$$C = C_c ,$$

(40)

$$D = D_c .$$

(41)

4. SIMULATION SYSTEM

This section depicts the computer simulation of the failure prediction for PM on a DC motor.

4.1 Parameters

Parameters for the DC motor in this simulation are prescribed as follows [20]:

E=10 V, B=0.001 N·m·sec, J=0.01 Kg·m², K_T=1 N·m / A, K_b=0.02 V·sec, R=10 Ω, L_a=0.01 H.

Substituting them into (35) and (36), the continuous state equations of the motor become

$$\frac{d}{dt}\begin{bmatrix} \theta \\ \dot{\theta} \\ i_a \end{bmatrix} = \begin{bmatrix} 0 & 1 & 0 \\ 0 & -0.1 & 100 \\ 0 & -2 & -1000 \end{bmatrix}\begin{bmatrix} \theta \\ \dot{\theta} \\ i_a \end{bmatrix} + \begin{bmatrix} 0 \\ 0 \\ 100 \end{bmatrix}10, \tag{42}$$

$$Y = \begin{bmatrix} 0 & 1 & 0 \end{bmatrix}\begin{bmatrix} \theta \\ \dot{\theta} \\ i_a \end{bmatrix}. \tag{43}$$

Besides, the following parameters are used to conduct failure prediction:

1. The failure threshold of the motor is defined as 5% less than the normal value, which is set to be the initial estimate in the Kalman prediction procedure. That is, the motor is judged to fail if the rotating speed drops to 95% of the normal value.

2. Mean time between failure (MTBF) for the motor is 100,000 hours [21].

3. Sampling interval T is 1 hour that is the increment time for every step in Kalman prediction.

4. Disturbance W_k has mean 0 and variance 0.01 V [22].

5. Measurement error V_k for θ has zero mean and standard deviation of 3.333 rad/sec, which is 1% full-scale accuracy [23] of the measurement.

6. PM lead-time is set at n × 60 minutes, where n is the ahead-step number for prediction. Accordingly, the alarm signal goes on for reminding PM to be executed whenever the Kalman filter predicts that the motor speed will be lower than the prescribed threshold n × 60 minutes later.

4.2 Monte Carlo Simulation and ARMA Model

Assuming failures of the motor occur randomly. Monte Carlo simulation (MCS) is adopted to generate failure times of the motor. The relation between failure rate $h(t)$ and distribution function of life $f(t)$ is [5]

$$f(t) = h(t)\exp[-\int_0^t h(\tau)d\tau]. \tag{44}$$

Failures occur randomly during the useful life period of a bathtub curve [5]. The failure rate is constant during this period. Let the failure rate in (44) be a constant λ, and (44) becomes

$$f(t) = \lambda\exp[-\int_0^t \lambda d\tau] = \lambda e^{-\lambda t}, \tag{45}$$

which is an exponential distribution function. Let u_i, i = 1, 2, 3, …, m, represent a set of standard uniformly distributed random numbers, the corresponding numbers t_i of the random variable t in (45), i.e. simulated failure times, are written as [5]

$$t_i = -\frac{1}{\lambda}\ln u_i, \tag{46}$$

with exponential distribution.

The measured data necessary for the recursive estimation loop of the Kalman filter, as depicted in Fig. 4, are generated by ARMA model, i.e. (1) to (3). Simulations in this section are performed by using MATALB [24]. All needed random numbers and white sequences with prescribed variances are obtained using the random number generator in MATLAB.

4.3 Exponential Attenuator

To account for the aging failure modes and the exponentially distributed failure times t_i, an exponential attenuator, represented as $e^{-t/\tau}$, is placed at output end of both motor system and the Kalman filter. The block diagram of simulation system is shown in Fig. 7. The symbol τ of the attenuator in Fig. 7 denotes the failure time constant of the motor, which varies with failure times that are generated by MCS.

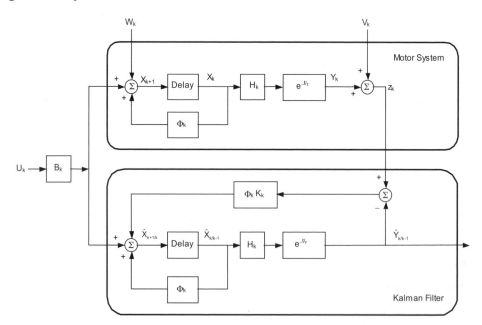

Figure 7.　Block diagram of the simulation system

4.4 Simulation Results

Two categories of simulation are conducted in this section, namely one-step-ahead prediction and two-step-ahead prediction. According to the central limit theorem (CLT), estimators follow the normal distribution if the sample size is sufficiently large. The sample size of 30 is a reasonable number to use [25]. The larger the sample size is, the smaller estimated error becomes, which tends to zero when the sample size approaches infinity. Hence, each simulation is decided to be executed 100 times. Simulation results for 60 minutes lead-time, i.e. one-step-ahead prediction, is shown in Fig. 8. Figure 8(a) shows the results of the 100 simulations of failure times generated by MCS, failure times predicted by Kalman filter, and the associated alarm times. Figure 8(b) shows the results of one of the 100 simulations with properly scaled coordinates. The failure time differences between MCS and Kalman prediction are shown in Fig. 9. The mean value and the standard deviation of the differences for the 100 simulations are -34.71 min and 65.90 min, respectively. The negative sign of the mean value indicates that the failure time predicted by Kalman filter is prior to the time generated by MCS. According to the Z formula [25], the error for estimating the mean value of the sample population can be calculated by

$$E_r^2 = \frac{Z_{\alpha/2}^2 \sigma^2}{n}.$$

The Z value for a 99% confidence level is 2.575 [25]. Solving for E_r gives

$$E_r = \frac{(2.575)(65.8954)}{\sqrt{100}} = 16.97(\text{min}).$$

According to the above data, there is 99% confidence to say that the interval for the mean value of the time difference between MCS and Kalman prediction is -34.71±16.97 min, i.e. from -17.74 min to -51.68 min. Taking the time difference into account, the alarm signal will appear at least 77.74 min prior to failure occurrence.

Results for the second category simulation, i.e. two-step-ahead prediction and lead-time for PM is 120 minutes, are shown in Figs. 10 and 11. The mean value of the failure time differences between MCS and Kalman prediction is -56.34 min, and the 99% confidence interval for this mean is 20.06 min. The maximum prediction error for this case is 76.40 min, which is 1.48 times greater than the error of the one-step-ahead prediction.

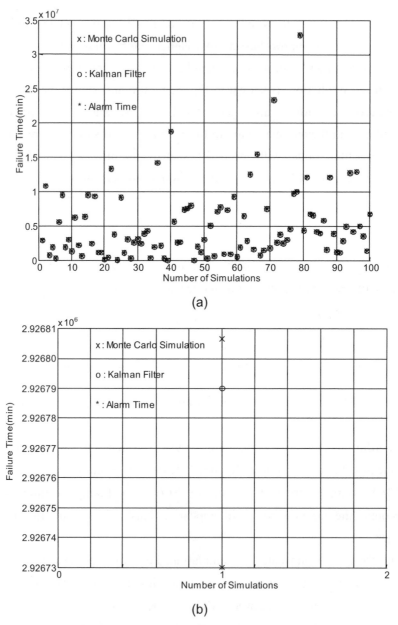

(a)

(b)

Figure 8. Failure time generated by Monte Carlo simulation and predicted by Kalman filter when lead-time=60 min

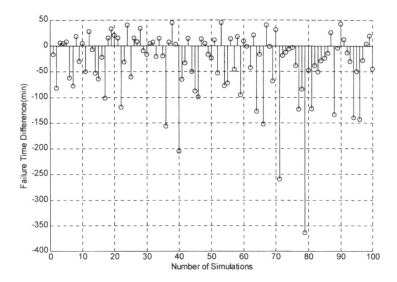

Figure 9.   Failure time difference between Monte Carlo simulation result
and Kalman filter prediction when lead-time=60 min

4.5   Discussions for the Simulation

1. In order to avoid false alarm, the failure threshold can not be set too close to the normal value. Otherwise, a decision-making algorithm is needed to identify that a failure indeed occurs.

2. The disturbance amplitude should be composed of all possible uncertainties of the motor and the environment.

3. The proposed method cannot deal with abrupt changes during a sampling interval. Thus, the sampling interval should not be too long.

4. Since the prediction is for PM purpose, the prediction time should be reasonably long enough for the PM action.

5. In contrast to the deterministic portion, the variance that is driven by the disturbance of the system is small. The difference of state variables between prediction steps fades very fast. Thus, using the N-step predictor, i.e. (20), only prediction result of the first several steps is of significance.

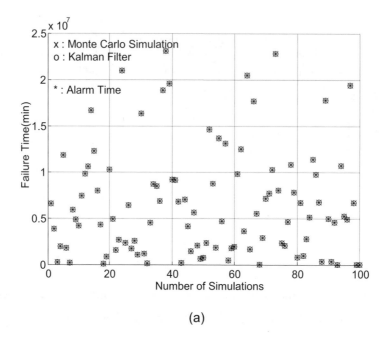

(a)

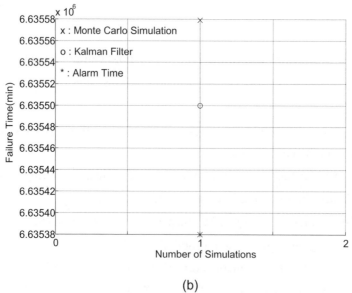

(b)

Figure 10.   Failure time generated by Monte Carlo simulation and
predicted by Kalman filter when lead-time=120 min

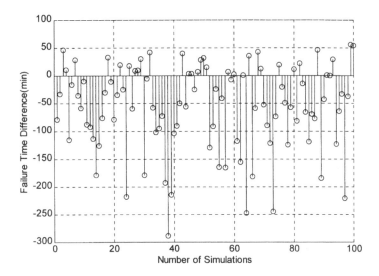

Figure 11. Failure time difference between Monte Carlo simulation result
and Kalman filter prediction when lead-time=120 min

6. The proposed method in this section is exemplified by a motor system, which
    is treated as a component. The procedure can be executed on a
    multi-component system if state equations for the components as a whole can
    be constructed. Performing the procedure on either the multi-component
    system or each of the components are both feasible. For a complicated or large
    system, the proposed method can be only performed on those elements in
    minimum cut sets that are constructed by fault tree analysis or Petri net model
    for failure [26].

7. Regarding multiple failure modes, they can be modeled to become modules,
    such as an attenuator for simulating aging failure mode for an electrical motor
    exemplified in this paper, and placed at the system model output end to extend
    the proposed method. As depicted previously, the system model may be
    single-component or multi-component. Whether the failure modules are
    placed in serial, parallel or other forms can be determined by system failure
    analysis [26]. As for a multi-component system with multiple failure modes,

the system can be taken apart to several components and placed the related failure module(s) at the output end of each component to perform state estimation by Kalman filter for each component.

5. EXPERIMENT ON AN ARMATURE-CONTROLLED DC MOTOR

This section presents the failure-prediction experiment for PM on a DC motor.

5.1 Experiment Setup

The experiment setup, as shown in Fig. 12, is composed of a DC motor with driver unit and a data acquisition system.

(a)

(b)

Figure 12.   Experiment setup

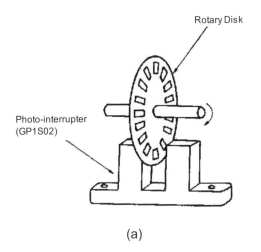

(a)

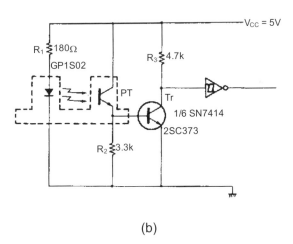

(b)

Figure 13.   Device and circuit for rotating speed measurement

DC Motor

The DC motor used in this experiment is made by TECO, Taiwan. The model number of the motor is GSDT-1/2 hp. Parameters for the DC motor used in this study are as follows [27]:

E=150 V, B=0.001135 N·m·sec, J=0.0102 kg·m², K_T=0.153 N·m / A,

K_b=1.926 V·sec, R=3.84 Ω, L_a=0.01 H.

Substituting them into (35) and (36), the continuous state equations of the motor become

$$\frac{d}{dt}\begin{bmatrix} \theta \\ \dot{\theta} \\ i_a \end{bmatrix} = \begin{bmatrix} 0 & 1 & 0 \\ 0 & -0.111 & 15 \\ 0 & -192.6 & -384 \end{bmatrix}\begin{bmatrix} \theta \\ \dot{\theta} \\ i_a \end{bmatrix} + \begin{bmatrix} 0 \\ 0 \\ 100 \end{bmatrix}150, \tag{47}$$

$$Y = \begin{bmatrix} 0 & 1 & 0 \end{bmatrix}\begin{bmatrix} \theta \\ \dot{\theta} \\ i_a \end{bmatrix}. \tag{48}$$

The discrete state equations sampled from (47) and (48) with time interval T=1200 seconds are

$$\begin{bmatrix} \theta_{k+1} \\ \dot{\theta}_{k+1} \\ i_{a,k+1} \end{bmatrix} = \begin{bmatrix} 1 & 0.13098 & 0.0051164 \\ 0 & 0 & 0 \\ 0 & 0 & 0 \end{bmatrix}\begin{bmatrix} \theta_k \\ \dot{\theta}_k \\ i_{a,k} \end{bmatrix} + \begin{bmatrix} 613.91 \\ 0.51164 \\ 0.0037955 \end{bmatrix}150, \tag{49}$$

$$Y_k = \begin{bmatrix} 0 & 1 & 0 \end{bmatrix}\begin{bmatrix} \theta_k \\ \dot{\theta}_k \\ i_{a,k} \end{bmatrix}. \tag{50}$$

Besides, the following parameters are used to conduct state estimation in this experiment:

1. Sampling interval T=20 minutes that is the increment time for every step in Kalman prediction. For comparing results among shorter and longer Ts, this study performed another two estimations with different time-intervals between two states, i.e. T=5 minutes and T=60 minutes.

2. Disturbance W_k has mean 0 and variance 0.1 V [22].

3. Measurement error V_k for $\dot{\theta}$ has zero mean and 1% full-scale accuracy [23] of the measurement.

4.   The rated rotating speed of the DC motor is 3180 rpm [27], which is prescribed as the initial value of the state variable θ.

Data Acquisition System

The data acquisition system used in this experiment is composed of a photo-interrupter circuit, a personal computer (PC), and a RS-232 transmission interface [23]. The rotating speed of the DC motor is measured by the photo-interrupter coded GP1S02. The shaft of a rotary disk is connected to the shaft of the DC motor, and the disk is placed between the light-emitting element and the light-receiving element of the photo-interrupter so as to generate pulse-signals while the motor rotates. The device and the circuit are shown in Fig. 13.

Pulse-signals are transmitted to the PC through the RS-232 interface, and the PC counts the pulses that are accumulated within 60 seconds in order to derive the rotating speed in rpm (revolution per minute).

5.2   Experiment Results

Results of the experiment are presented and discussed in this section.

Measured Data

The rotating speed of the motor was measured and recorded per 5 minutes day and night for 80 days. Because the experiment lasts for nearly three months, a large number of data were thus accumulated. There are 288 measurements in a day and 23040 data in total for that period of time. Figure 14 shows the results. The data were fed into the estimator, as depicted in Fig. 4, to estimate the one-step-ahead state variables. The measured data and the resultant estimates for T=20 minutes (i.e. every four measurements) are shown in Fig. 15. The data will be mixed up and become hard to read if all 23040 data are shown in one chart. To avoid this and to present the results more clearly, the time-axis unit of Fig. 15 is set to be 24 hours (i.e. one point per day).

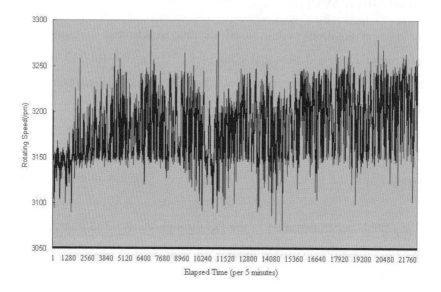

Figure 14. Measured rotating speed for every 5 minutes

Estimate Error Percentage

The estimate error percentage is defined as

$$E_r\% = \frac{\hat{\dot{\theta}}_{k+1/k} - \dot{\theta}_{k+1}}{\dot{\theta}_{k+1}} \times 100\% .$$ (51)

E_r represents the difference between predicted value and actual value. Fig. 16 shows the results that are derived from (51) using the data in Fig. 15. Reading from Fig. 16, the maximum $E_r\%$ is less than 3%.

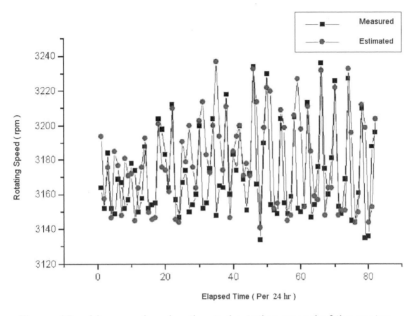

Figure 15.   Measured and estimated rotating speed of the motor

Mean Value and Variance of the Estimate Accuracy

Let

$$X_i = 1 - E_r\%_i \qquad i = 1, 2, \ldots, 23040 \tag{52}$$

be the individual accuracy of each estimate, and

$$\mu = \frac{\displaystyle\sum_{i=1}^{23040} X_i}{23040}, \tag{53}$$

$$\sigma^2 = \frac{\displaystyle\sum_{i=1}^{23040} (X_i - \mu)^2}{23040} \tag{54}$$

be the mean value and the variance [25] of the accuracy for the 23040 samples, respectively. According to (52), (53), and (54), the resultant mean values, variances, and standard deviations (σ) of the estimate accuracy for T=5, 20, 60 minutes are summarized in Table 1.

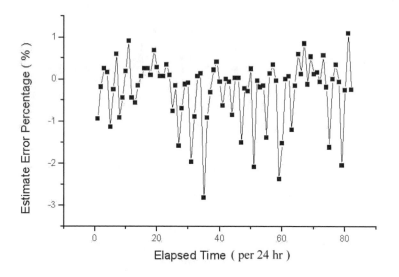

Figure 16.   Estimate Error percentage

Table 1.   Mean value, standard deviation, and variance for different T

T(minutes)	$\mu(\%)$	$\sigma(\%)$	$\sigma^2(\%)$
5	99.74656	0.402957	0.162374
20	99.74060	0.471612	0.222418
60	99.72771	0.652469	0.425716

Rotating Speed Variation

Variation percentage of the estimated rotating speed of the DC motor is defined as

$$V\% = \frac{\hat{\theta}_{k+1/k} - 3180}{3180} \times 100\% . \tag{55}$$

$V\%$ represents the variation percentage of the estimated rotating speed from the rated value 3180 rpm, i.e. the abnormality extent of the motor performance. It is used to judge whether the motor is going to fail or not. Since the MTBF of a motor is about 100,000 hours [21], the rotating speed of the motor in this experiment varied less than 2% of the rated value during the experiment time period. Variation percentage of the estimated rotating speed of the DC motor is shown in Fig. 17.

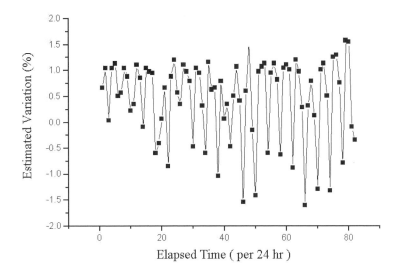

Figure 17. Variation percentage of the estimated rotating speed

5.3 Discussions for the Experiment

1. The mean estimate accuracies for T=5, 20, and 60 minutes are all higher than 99.7%, which infers that the one-step-ahead state variable can be accurately predicted using the proposed method in this experiment.
2. A threshold is a value used to judge an equipment failure occurs or not. It is prescribed as the measurement value that is taken just prior to or at the time of failure [28]. For failure prediction, the threshold for a motor should be determined by the user of the motor according to requirements for specific

situations. Once the estimated value reaches the threshold, the failure is predicted.

3. The disturbance amplitude should be composed of all possible uncertainties of the motor and the environment.

4. Since the prediction is for PM purpose, the prediction time should be reasonably long enough for the PM action.

5. The proposed method in this study is exemplified by a motor system, which is treated as a component. The procedure can be executed on a multi-component system if state equations for the components as a whole can be constructed. Performing the procedure on either the multi-component system or each of the components are both feasible. For a complicated or large system, the proposed method can be performed on those elements in minimum cut sets that are constructed by fault tree analysis or Petri net model for failure [26].

6. CONCLUSIONS

Knowing when and where a system needs maintenance and economizing capital investment are two of the major problems of maintenance. The proposed scheme improves the maintenance problem in the following aspects.

1. Before a system failure occurs, the scheme is able to indicate where and when the failure is going to be.

2. It makes the health condition and the historical record of maintenance for a system clear at a glance.

3. Scheduled maintenance is enacted based on a statistical average, which still retains the unavoidable risk that the system may fail before criteria are exceeded, i.e. a failure may occur unexpectedly. On the other hand, the actual duty-cycles for a certain part or module may be longer than those averages, so if they are replaced during scheduled maintenance, that is a waste of the investment. The condition-based scheme avoids those drawbacks.

Failure prediction simulation and experiment for PM by state estimation through Kalman filtering has been performed in this chapter. In contrast to previous works, this study uses the Kalman filtering instead of parameter trend to predict the time of failure occurrence and to determine the PM execution time. Resultant prediction errors of the simulation are acceptable not only for one-step-ahead prediction but also for two-step-ahead prediction. To simulate the aging failure mode, a state variable, i.e. rotating speed, is monitored in the simulation. The more variables are measured, the more complicated failure modes can be simulated. Moreover, an experiment of state estimation for predictive maintenance using the Kalman filter on a DC motor has also been performed in this chapter. Resultant prediction errors for one-step-ahead prediction are acceptable. Besides, the shorter the increment time for every step in Kalman prediction uses, the higher prediction accuracy it achieves. Considerations for determining the required PM lead-time and the increment time for prediction contradict to each other. How to compromise them and end up with an optimal value is important. Incorporating with fault tree analysis or Petri net model for failure, the proposed method can be only performed on those elements in minimum cut sets of a complicated or large system instead of on all elements of the whole system. Failure can be prevented in time so as to promote reliability by state estimation for predictive maintenance using the Kalman filter.

ACKNOWLEDGEMENT

This chapter quotes the contents of following papers with permission from Elsevier:

1. Yang, S. K. and Liu, T. S., 'State estimation for predictive maintenance using Kalman filter', *Reliability Engineering and System Safety*, **66**(1), pp. 29-39, October 1999.

2. Yang, S. K., 'An experiment of state estimation for predictive maintenance using Kalman filter on a DC motor', *Reliability Engineering and System Safety*, **75**(1), pp. 103-111, January 2002.

REFERENCES

[1] E. A. Elsayed, *Reliability Engineering*: Addison Wesley Longman, 1996.

[2] J. D. Patton, Jr., *Preventive Maintenance*: Instrument Society of America, 1983.

[3] IEC 50(191), *International Electrotechnical Vocabulary (IEV)*, Chapter 191—Dependability and quality of service, International Electrotechnical Commission, Geneva, 1990.

[4] M. Rausand and K. Oien, 'The basic concept of failure analysis', *Reliability Engineering and System Safety*, **53**, 73-83 (1996).

[5] S. S. Rao, *Reliability-Based Design*: McGraw-Hill, 1992.

[6] O. T. Ogunyemi and P. I. Nelson, "Prediction of Gamma failure times," *IEEE Transactions on Reliability*, vol. 46, No. 3, pp. 400-405, 1997.

[7] W. Nelson, "Weibull prediction of a future number of failures," *Quality and Reliability Engineering International*, vol. 16, pp. 23-26, 2000.

[8] E. A. Rietman and M. Beachy, "A study on failure prediction in a plasma reactor," *IEEE Transactions on Semiconductor Manufacturing*, vol. 11, No. 4, pp. 670-680, 1998.

[9] S. N. Kher and G. M. Bubel, "Predicting system-failure risk from unanticipated fiber-breaks in manufacturing," *IEEE Transactions on Reliability*, vol. 47, No. 2, pp. 126-130, 1998.

[10] R. Isermann, "Process fault detection based on modeling and estimation methods—A survey," *Automatica*, vol. 20, no. 4, pp. 387-404, 1984.

[11] J. L. Tylee, "On-line failure detection in nuclear power plant instrumentation", *IEEE Transactions on Automatic Control*, vol. AC-28, no. 3, pp. 406-415, 1983.

[12] M. M. Sidar and B. F. Doolin, "On th feasibility of real-time prediction of aircraft carrier motion at sea", *IEEE Transactions on Automatic Control*, vol. AC-28, no. 3, pp. 350-355, 1983.

[13] R. F. Berg, "Estimation and prediction for maneuvering target trajectories", *IEEE Transactions on Automatic Control*, vol. AC-28, no. 3, pp. 294-304, 1983.

[14] B. D. O. Anderson and J. B. Moore, *Optimal Filtering*: Prentice-Hall, 1979.

[15] A. S. Willsky, "A survey of design methods for failure detection in dynamic systems," *Automatica*, vol. 12, pp. 601-611, 1976.

[16] R. G. Brown and P. Y. C. Hwang, *Introduction to Random Signals and Applied Kalman Filtering*: John Wiley & Sons, Sec. 5.3-5.6, pp. 198-224, 1997.

[17] K. Ogata, *Modern Control Engineering*, Prentice-Hall, Englewood Cliffs, New Jersey, 1980.

[18] C. L. Phillips and H. T. Nagle, *Digital Control System Analysis and Design*, Prentice-Hall, Englewood Cliffs, New Jersey, 1990.

[19] K. Ogata, *Discrete-Time Control Systems*, Prentice-Hall, Englewood Cliffs, New Jersey, 1987.

[20] G. F. Franklin, J. D. Powell and A. Emami-Naeini, *Feedback Control of Dynamic Systems*, Addison-Wesley, New York, 1994.

[21] B. S. Dhillon, *Mechanical Reliability: Theory, Models and Applications*, AIAA Education Series, Washington DC, 1988.

[22] C. Y. Chiang-Lin, "Parameter estimation and fault diagnosis on the thermal network by applying the extended Kalman filter and the expert system", *M.S.-thesis*, National Chiao Tung University, Taiwan, Republic of China, 1991.

[23] J. G. Webster and R. Palla's-Areny, *Sensor and Signal Conditioning*, John Wiley & Sons, New York, 1991.

[24] D. Fowley and M. Horton, *User's Guide of MATLAB*, Version 4, Prentice-Hall, Englewood Cliffs, New Jersey, 1995.

[25] K. Black, *Business Statistics*, *Contemporary Decision Making*, West Publishing, Los Angeles, 1997.

[26] S. K. Yang and T. S. Liu, "Failure analysis for an airbag inflator by Petri nets," *Quality and Reliability Engineering International*, vol. 13, pp. 139-151, 1997.

[27] TECO, DC motor-Universal type User Guide. Taipei: Tung-yuang Electrical machine Company, 2001.

[28] S. K. Yang and T. S. Liu, "A Petri net approach to early failure detection and isolation for preventive maintenance," *Quality and Reliability Engineering International*, vol. 14, pp. 319-330, 1998.

附錄 F

標準常態分配值表*

Normal Distribution and Related Functions

$$F(x) = \int_{-\infty}^{x} \frac{1}{\sqrt{2\pi}}\, e^{-\frac{1}{2}t^2}\, dt$$

$$f(x) = \frac{1}{\sqrt{2\pi}}\, e^{-\frac{1}{2}x^2}$$

註：本表所用之符號「x」，即爲課文中(例：2.5 節，標準常態分配)的「Z」。

x	$F(x)$	$1 - F(x)$	$f(x)$	x	$F(x)$	$1 - F(x)$	$f(x)$
.00	.5000	.5000	.3989	.51	.6950	.3050	.3503
.01	.5040	.4960	.3989	.52	.6985	.3015	.3485
.02	.5080	.4920	.3989	.53	.7019	.2981	.3467
.03	.5120	.4880	.3988	.54	.7054	.2946	.3448
.04	.5160	.4840	.3986	.55	.7088	.2912	.3429
.05	.5199	.4801	.3984	.56	.7123	.2877	.3410
.06	.5239	.4761	.3982	.57	.7157	.2843	.3391
.07	.5279	.4721	.3980	.58	.7190	.2810	.3372
.08	.5319	.4681	.3977	.59	.7224	.2776	.3352
.09	.5359	.4641	.3973	.60	.7257	.2743	.3332
.10	.5398	.4602	.3970	.61	.7291	.2709	.3312
.11	.5438	.4562	.3965	.62	.7324	.2676	.3292
.12	.5478	.4522	.3961	.63	.7357	.2643	.3271
.13	.5517	.4483	.3956	.64	.7389	.2611	.3251
.14	.5557	.4443	.3951	.65	.7422	.2578	.3230
.15	.5596	.4404	.3945	.66	.7454	.2546	.3209
.16	.5636	.4364	.3939	.67	.7486	.2514	.3187
.17	.5675	.4325	.3932	.68	.7517	.2483	.3166
.18	.5714	.4286	.3925	.69	.7549	.2451	.3144
.19	.5753	.4247	.3918	.70	.7580	.2420	.3123
.20	.5793	.4207	.3910	.71	.7611	.2389	.3101
.21	.5832	.4168	.3902	.72	.7642	.2358	.3079
.22	.5871	.4129	.3894	.73	.7673	.2327	.3056
.23	.5910	.4090	.3885	.74	.7704	.2296	.3034
.24	.5948	.4052	.3876	.75	.7734	.2266	.3011
.25	.5987	.4013	.3867	.76	.7764	.2236	.2989
.26	.6026	.3974	.3857	.77	.7794	.2206	.2966
.27	.6064	.3936	.3847	.78	.7823	.2177	.2943
.28	.6103	.3897	.3836	.79	.7852	.2148	.2920
.29	.6141	.3859	.3825	.80	.7881	.2119	.2897
.30	.6179	.3821	.3814	.81	.7910	.2090	.2874
.31	.6217	.3783	.3802	.82	.7939	.2061	.2850
.32	.6255	.3745	.3790	.83	.7967	.2033	.2827
.33	.6293	.3707	.3778	.84	.7995	.2005	.2803
.34	.6331	.3669	.3765	.85	.8023	.1977	.2780
.35	.6368	.3632	.3752	.86	.8051	.1949	.2756
.36	.6406	.3594	.3739	.87	.8078	.1922	.2732
.37	.6443	.3557	.3725	.88	.8106	.1894	.2709
.38	.6480	.3520	.3712	.89	.8133	.1867	.2685
.39	.6517	.3483	.3697	.90	.8159	.1841	.2661
.40	.6554	.3446	.3683	.91	.8186	.1814	.2637
.41	.6591	.3409	.3668	.92	.8212	.1788	.2613
.42	.6628	.3372	.3653	.93	.8238	.1762	.2589
.43	.6664	.3336	.3637	.94	.8264	.1736	.2565
.44	.6700	.3300	.3621	.95	.8289	.1711	.2541

x	F(x)	1 - F(x)	f(x)	x	F(x)	1 - F(x)	f(x)
.45	.6736	.3264	.3605	.96	.8315	.1685	.2516
.46	.6772	.3228	.3589	.97	.8340	.1660	.2492
.47	.6808	.3192	.3572	.98	.8365	.1635	.2468
.48	.6844	.3156	.3555	.99	.8389	.1611	.2444
.49	.6879	.3121	.3538	1.00	.8413	.1587	.2420
.50	.6915	.3085	.3521	1.01	.8438	.1562	.2396
1.02	.8461	.1539	.2371	1.51	.9345	.0655	.1276
1.03	.8485	.1515	.2347	1.52	.9357	.0643	.1257
1.04	.8508	.1492	.2323	1.53	.9370	.0630	.1238
1.05	.8531	.1469	.2299	1.54	.9382	.0618	.1219
1.06	.8554	.1446	.2275	1.55	.9394	.0606	.1200
1.07	.8577	.1423	.2251	1.56	.9406	.0594	.1182
1.08	.8599	.1401	.2227	1.57	.9418	.0582	.1163
1.09	.8621	.1379	.2203	1.58	.9429	.0571	.1145
1.10	.8643	.1357	.2179	1.59	.9441	.0559	.1127
1.11	.8665	.1335	.2155	1.60	.9452	.0548	.1109
1.12	.8686	.1314	.2131	1.61	.9463	.0537	.1092
1.13	.8708	.1292	.2107	1.62	.9474	.0526	.1074
1.14	.8729	.1271	.2083	1.63	.9484	.0516	.1057
1.15	.8749	.1251	.2059	1.64	.9495	.0505	.1040
1.16	.8770	.1230	.2036	1.65	.9505	.0495	.1023
1.17	.8790	.1210	.2012	1.66	.9515	.0485	.1006
1.18	.8810	.1190	.1989	1.67	.9525	.0475	.0989
1.19	.8830	.1170	.1965	1.68	.9535	.0465	.0973
1.20	.8849	.1151	.1942	1.69	.9545	.0455	.0957
1.21	.8869	.1131	.1919	1.70	.9554	.0446	.0940
1.22	.8888	.1112	.1895	1.71	.9564	.0436	.0925
1.23	.8907	.1093	.1872	1.72	.9573	.0427	.0909
1.24	.8925	.1075	.1849	1.73	.9582	.0418	.0893
1.25	.8944	.1056	.1826	1.74	.9591	.0409	.0878
1.26	.8962	.1038	.1804	1.75	.9599	.0401	.0863
1.27	.8980	.1020	.1781	1.76	.9608	.0392	.0848
1.28	.8997	.1003	.1758	1.77	.9616	.0384	.0833
1.29	.9015	.0985	.1736	1.78	.9625	.0375	.0818
1.30	.9032	.0968	.1714	1.79	.9633	.0367	.0804
1.31	.9049	.0951	.1691	1.80	.9641	.0359	.0790
1.32	.9066	.0934	.1669	1.81	.9649	.0351	.0775
1.33	.9082	.0918	.1647	1.82	.9656	.0344	.0761
1.34	.9099	.0901	.1626	1.83	.9664	.0336	.0748
1.35	.9115	.0885	.1604	1.84	.9671	.0329	.0734
1.36	.9131	.0869	.1582	1.85	.9678	.0322	.0721
1.37	.9147	.0853	.1561	1.86	.9686	.0314	.0707
1.38	.9162	.0838	.1539	1.87	.9693	.0307	.0694
1.39	.9177	.0823	.1518	1.88	.9699	.0301	.0681

x	$F(x)$	$1 - F(x)$	$f(x)$	x	$F(x)$	$1 - F(x)$	$f(x)$
1.40	.9192	.0808	.1497	1.89	.9706	.0294	.0669
1.41	.9207	.0793	.1476	1.90	.9713	.0287	.0656
1.42	.9222	.0778	.1456	1.91	.9719	.0281	.0644
1.43	.9236	.0764	.1435	1.92	.9726	.0274	.0632
1.44	.9251	.0749	.1415	1.93	.9732	.0268	.0620
1.45	.9265	.0735	.1394	1.94	.9738	.0262	.0608
1.46	.9279	.0721	.1374	1.95	.9744	.0256	.0596
1.47	.9292	.0708	.1354	1.96	.9750	.0250	.0584
1.48	.9306	.0694	.1334	1.97	.9756	.0244	.0573
1.49	.9319	.0681	.1315	1.98	.9761	.0239	.0562
1.50	.9332	.0668	.1295	1.99	.9767	.0233	.0551
2.00	.9772	.0228	.0540	2.51	.9940	.0060	.0717
2.01	.9778	.0222	.0529	2.52	.9941	.0059	.0167
2.02	.9783	.0217	.0519	2.53	.9943	.0057	.0163
2.03	.9788	.0212	.0508	2.54	.9945	.0055	.0158
2.04	.9793	.0207	.0498	2.55	.9946	.0054	.0155
2.05	.9798	.0202	.0488	2.56	.9948	.0052	.0151
2.06	.9803	.0197	.0478	2.57	.9949	.0051	.0147
2.07	.9808	.0192	.0468	2.58	.9951	.0049	.0143
2.08	.9812	.0188	.0459	2.59	.9952	.0048	.0139
2.09	.9817	.0183	.0449	2.60	.9953	.0047	.0136
2.10	.9821	.0179	.0440	2.61	.9955	.0045	.0132
2.11	.9826	.0174	.0431	2.62	.9956	.0044	.0129
2.12	.9830	.0170	.0422	2.63	.9957	.0043	.0126
2.13	.9834	.0166	.0413	2.64	.9959	.0041	.0122
2.14	.9838	.0162	.0404	2.65	.9960	.0040	.0119
2.15	.9842	.0158	.0396	2.66	.9961	.0039	.0116
2.16	.9846	.0154	.0387	2.67	.9962	.0038	.0113
2.17	.9850	.0150	.0379	2.68	.9963	.0037	.0110
2.18	.9854	.0146	.0371	2.69	.9964	.0036	.0107
2.19	.9857	.0143	.0363	2.70	.9965	.0035	.0104
2.20	.9861	.0139	.0355	2.71	.9966	.0034	.0101
2.21	.9864	.0136	.0347	2.72	.9967	.0033	.0099
2.22	.9868	.0132	.0339	2.73	.9968	.0032	.0096
2.23	.9871	.0129	.0332	2.74	.9969	.0031	.0093
2.24	.9875	.0125	.0325	2.75	.9970	.0030	.0091
2.25	.9878	.0122	.0317	2.76	.9971	.0029	.0088
2.26	.9881	.0119	.0310	2.77	.9972	.0028	.0086
2.27	.9884	.0116	.0303	2.78	.9973	.0027	.0084
2.28	.9887	.0113	.0297	2.79	.9974	.0026	.0081
2.29	.9890	.0110	.0290	2.80	.9974	.0026	.0079
2.30	.9893	.0107	.0283	2.81	.9975	.0025	.0077
2.31	.9896	.0104	.0277	2.82	.9976	.0024	.0075
2.32	.9898	.0102	.0270	2.83	.9977	.0023	.0073

x	F(x)	1 - F(x)	f(x)	x	F(x)	1 - F(x)	f(x)
2.33	.9901	.0099	.0264	2.84	.9977	.0023	.0071
2.34	.9904	.0096	.0258	2.85	.9978	.0022	.0069
2.35	.9906	.0094	.0252	2.86	.9979	.0021	.0067
2.36	.9909	.0091	.0246	2.87	.9979	.0021	.0065
2.37	.9911	.0089	.0241	2.88	.9980	.0020	.0063
2.38	.9913	.0087	.0235	2.89	.9981	.0019	.0061
2.39	.9916	.0084	.0229	2.90	.9981	.0019	.0060
2.40	.9918	.0082	.0224	2.91	.9982	.0018	.0058
2.41	.9920	.0080	.0219	2.92	.9982	.0018	.0056
2.42	.9922	.0078	.0213	2.93	.9983	.0017	.0055
2.43	.9925	.0075	.0208	2.94	.9984	.0016	.0053
2.44	.9927	.0073	.0203	2.95	.9984	.0016	.0051
2.45	.9929	.0071	.0198	2.96	.9985	.0015	.0050
2.46	.9931	.0069	.0194	2.97	.9985	.0015	.0048
2.47	.9932	.0068	.0189	2.98	.9986	.0014	.0047
2.48	.9934	.0066	.0184	2.99	.9986	.0014	.0046
2.49	.9936	.0064	.0180	3.00	.9987	.0013	.0044
2.50	.9938	.0062	.0175	3.01	.9987	.0013	.0043
3.02	.9987	.0013	.0042	3.51	.9998	.0002	.0008
3.03	.9988	.0012	.0040	3.52	.9998	.0002	.0008
3.04	.9988	.0012	.0039	3.53	.9998	.0002	.0008
3.05	.9989	.0011	.0038	3.54	.9998	.0002	.0008
3.06	.9989	.0011	.0037	3.55	.9998	.0002	.0007
3.07	.9989	.0011	.0036	3.56	.9998	.0002	.0007
3.08	.9990	.0010	.0035	3.57	.9998	.0002	.0007
3.09	.9990	.0010	.0034	3.58	.9998	.0002	.0007
3.10	.9990	.0010	.0033	3.59	.9998	.0002	.0006
3.11	.9991	.0009	.0032	3.60	.9998	.0002	.0006
3.12	.9991	.0009	.0031	3.61	.9998	.0002	.0006
3.13	.9991	.0009	.0030	3.62	.9999	.0001	.0006
3.14	.9992	.0008	.0029	3.63	.9999	.0001	.0005
3.15	.9992	.0008	.0028	3.64	.9999	.0001	.0005
3.16	.9992	.0008	.0027	3.65	.9999	.0001	.0005
3.17	.9992	.0008	.0026	3.66	.9999	.0001	.0005
3.18	.9993	.0007	.0025	3.67	.9999	.0001	.0005
3.19	.9993	.0007	.0025	3.68	.9999	.0001	.0005
3.20	.9993	.0007	.0024	3.69	.9999	.0001	.0004
3.21	.9993	.0007	.0023	3.70	.9999	.0001	.0004
3.22	.9994	.0006	.0022	3.71	.9999	.0001	.0004
3.23	.9994	.0006	.0022	3.72	.9999	.0001	.0004
3.24	.9994	.0006	.0021	3.73	.9999	.0001	.0004
3.25	.9994	.0006	.0020	3.74	.9999	.0001	.0004
3.26	.9994	.0006	.0020	3.75	.9999	.0001	.0004

x	F(x)	1 - F(x)	f(x)	x	F(x)	1 - F(x)	f(x)
3.27	.9995	.0005	.0019	3.76	.9999	.0001	.0003
3.28	.9995	.0005	.0018	3.77	.9999	.0001	.0003
3.29	.9995	.0005	.0018	3.78	.9999	.0001	.0003
3.30	.9995	.0005	.0017	3.79	.9999	.0001	.0003
3.31	.9995	.0005	.0017	3.80	.9999	.0001	.0003
3.32	.9995	.0005	.0016	3.81	.9999	.0001	.0003
3.33	.9996	.0004	.0016	3.82	.9999	.0001	.0003
3.34	.9996	.0004	.0015	3.83	.9999	.0001	.0003
3.35	.9996	.0004	.0015	3.84	.9999	.0001	.0003
3.36	.9996	.0004	.0014	3.85	.9999	.0001	.0002
3.37	.9996	.0004	.0014	3.86	.9999	.0001	.0002
3.38	.9996	.0004	.0013	3.87	.9999	.0001	.0002
3.39	.9997	.0003	.0013	3.88	.9999	.0001	.0002
3.40	.9997	.0003	.0012	3.89	1.0000	.0000	.0002
3.41	.9997	.0003	.0012	3.90	1.0000	.0000	.0002
3.42	.9997	.0003	.0012	3.91	1.0000	.0000	.0002
3.43	.9997	.0003	.0011	3.92	1.0000	.0000	.0002
3.44	.9997	.0003	.0011	3.93	1.0000	.0000	.0002
3.45	.9997	.0003	.0010	3.94	1.0000	.0000	.0002
3.46	.9997	.0003	.0010	3.95	1.0000	.0000	.0002
3.47	.9997	.0003	.0010	3.96	1.0000	.0000	.0002
3.48	.9997	.0003	.0009	3.97	1.0000	.0000	.0002
3.49	.9998	.0002	.0009	3.98	1.0000	.0000	.0001
3.50	.9998	.0002	.0009	3.99	1.0000	.0000	.0001
				4.00	1.0000	.0000	.0001

附錄 G

GAMMA 函數值表

Gamma Function

n	$\Gamma(n)$	n	$\Gamma(n)$	n	$\Gamma(n)$	n	$\Gamma(n)$
0.0100	99.4327	0.5100	1.7384	1.0100	0.9943	1.5100	0.8866
0.0200	49.4423	0.5200	1.7058	1.0200	0.9888	1.5200	0.8870
0.0300	32.7850	0.5300	1.6747	1.0300	0.9836	1.5300	0.8876
0.0400	24.4610	0.5400	1.6448	1.0400	0.9784	1.5400	0.8882
0.0500	19.4701	0.5500	1.6161	1.0500	0.9735	1.5500	0.8889
0.0600	16.1457	0.5600	1.5886	1.0600	0.9687	1.5600	0.8896
0.0700	13.7736	0.5700	1.5623	1.0700	0.9642	1.5700	0.8905
0.0800	11.9966	0.5800	1.5369	1.0800	0.9597	1.5800	0.8914
0.0900	10.6162	0.5900	1.5126	1.0900	0.9555	1.5900	0.8924
0.1000	9.5135	0.6000	1.4892	1.1000	0.9513	1.6000	0.8935
0.1100	8.6127	0.6100	1.4667	1.1100	0.9474	1.6100	0.8947
0.1200	7.8632	0.6200	1.4450	1.1200	0.9436	1.6200	0.8959
0.1300	7.2302	0.6300	1.4242	1.1300	0.9399	1.6300	0.8972
0.1400	6.6887	0.6400	1.4041	1.1400	0.9364	1.6400	0.8986
0.1500	6.2203	0.6500	1.3848	1.1500	0.9330	1.6500	0.9001
0.1600	5.8113	0.6600	1.3662	1.1600	0.9298	1.6600	0.9017
0.1700	5.4512	0.6700	1.3482	1.1700	0.9267	1.6700	0.9033
0.1800	5.1318	0.6800	1.3309	1.1800	0.9237	1.6800	0.9050
0.1900	4.8468	0.6900	1.3142	1.1900	0.9209	1.6900	0.9068
0.2000	4.5908	0.7000	1.2981	1.2000	0.9182	1.7000	0.9086
0.2100	4.3599	0.7100	1.2825	1.2100	0.9156	1.7100	0.9106
0.2200	4.1505	0.7200	1.2675	1.2200	0.9131	1.7200	0.9126
0.2300	3.9598	0.7300	1.2530	1.2300	0.9108	1.7300	0.9147
0.2400	3.7855	0.7400	1.2390	1.2400	0.9085	1.7400	0.9168
0.2500	3.6256	0.7500	1.2254	1.2500	0.9064	1.7500	0.9191
0.2600	3.4785	0.7600	1.2123	1.2600	0.9044	1.7600	0.9214
0.2700	3.3426	0.7700	1.1997	1.2700	0.9025	1.7700	0.9238
0.2800	3.2169	0.7800	1.1875	1.2800	0.9007	1.7800	0.9262
0.2900	3.1001	0.7900	1.1757	1.2900	0.8990	1.7900	0.9288
0.3000	2.9916	0.8000	1.1642	1.3000	0.8975	1.8000	0.9314
0.3100	2.8903	0.8100	1.1532	1.3100	0.8960	1.8100	0.9341
0.3200	2.7958	0.8200	1.1425	1.3200	0.8946	1.8200	0.9368
0.3300	2.7072	0.8300	1.1322	1.3300	0.8934	1.8300	0.9397
0.3400	2.6242	0.8400	1.1222	1.3400	0.8922	1.8400	0.9426
0.3500	2.5461	0.8500	1.1125	1.3500	0.8912	1.8500	0.9456
0.3600	2.4727	0.8600	1.1031	1.3600	0.8902	1.8600	0.9487
0.3700	2.4036	0.8700	1.0941	1.3700	0.8893	1.8700	0.9518
0.3800	2.3383	0.8800	1.0853	1.3800	0.8885	1.8800	0.9551
0.3900	2.2765	0.8900	1.0768	1.3900	0.8879	1.8900	0.9584
0.4000	2.2182	0.9000	1.0686	1.4000	0.8873	1.9000	0.9618
0.4100	2.1628	0.9100	1.0607	1.4100	0.8868	1.9100	0.9652
0.4200	2.1104	0.9200	1.0530	1.4200	0.8864	1.9200	0.9688
0.4300	2.0605	0.9300	1.0456	1.4300	0.8860	1.9300	0.9724
0.4400	2.0132	0.9400	1.0384	1.4400	0.8858	1.9400	0.9761
0.4500	1.9681	0.9500	1.0315	1.4500	0.8857	1.9500	0.9799
0.4600	1.9252	0.9600	1.0247	1.4600	0.8856	1.9600	0.9837
0.4700	1.8843	0.9700	1.0182	1.4700	0.8856	1.9700	0.9877
0.4800	1.8453	0.9800	1.0119	1.4800	0.8857	1.9800	0.9917
0.4900	1.8080	0.9900	1.0059	1.4900	0.8859	1.9900	0.9958
0.5000	1.7725	1.0000	1.0000	1.5000	0.8862	2.0000	1.0000

Gamma Function

n	$\Gamma(n)$	n	$\Gamma(n)$	n	$\Gamma(n)$	n	$\Gamma(n)$
2.0100	1.0043	2.5100	1.3388	3.0100	2.0186	3.5100	3.3603
2.0200	1.0086	2.5200	1.3483	3.0200	2.0374	3.5200	3.3977
2.0300	1.0131	2.5300	1.3580	3.0300	2.0565	3.5300	3.4357
2.0400	1.0176	2.5400	1.3678	3.0400	2.0759	3.5400	3.4742
2.0500	1.0222	2.5500	1.3777	3.0500	2.0955	3.5500	3.5132
2.0600	1.0269	2.5600	1.3878	3.0600	2.1153	3.5600	3.5529
2.0700	1.0316	2.5700	1.3981	3.0700	2.1355	3.5700	3.5930
2.0800	1.0365	2.5800	1.4084	3.0800	2.1559	3.5800	3.6338
2.0900	1.0415	2.5900	1.4190	3.0900	2.1766	3.5900	3.6751
2.1000	1.0465	2.6000	1.4296	3.1000	2.1976	3.6000	3.7170
2.1100	1.0516	2.6100	1.4404	3.1100	2.2189	3.6100	3.7595
2.1200	1.0568	2.6200	1.4514	3.1200	2.2405	3.6200	3.8027
2.1300	1.0621	2.6300	1.4625	3.1300	2.2623	3.6300	3.8464
2.1400	1.0675	2.6400	1.4738	3.1400	2.2845	3.6400	3.8908
2.1500	1.0730	2.6500	1.4852	3.1500	2.3069	3.6500	3.9358
2.1600	1.0786	2.6600	1.4968	3.1600	2.3297	3.6600	3.9814
2.1700	1.0842	2.6700	1.5085	3.1700	2.3528	3.6700	4.0277
2.1800	1.0900	2.6800	1.5204	3.1800	2.3762	3.6800	4.0747
2.1900	1.0959	2.6900	1.5325	3.1900	2.3999	3.6900	4.1223
2.2000	1.1018	2.7000	1.5447	3.2000	2.4240	3.7000	4.1707
2.2100	1.1078	2.7100	1.5571	3.2100	2.4483	3.7100	4.2197
2.2200	1.1140	2.7200	1.5696	3.2200	2.4731	3.7200	4.2694
2.2300	1.1202	2.7300	1.5824	3.2300	2.4981	3.7300	4.3199
2.2400	1.1266	2.7400	1.5953	3.2400	2.5235	3.7400	4.3711
2.2500	1.1330	2.7500	1.6084	3.2500	2.5493	3.7500	4.4230
2.2600	1.1395	2.7600	1.6216	3.2600	2.5754	3.7600	4.4757
2.2700	1.1462	2.7700	1.6351	3.2700	2.6018	3.7700	4.5291
2.2800	1.1529	2.7800	1.6487	3.2800	2.6287	3.7800	4.5833
2.2900	1.1598	2.7900	1.6625	3.2900	2.6559	3.7900	4.6384
2.3000	1.1667	2.8000	1.6765	3.3000	2.6834	3.8000	4.6942
2.3100	1.1738	2.8100	1.6907	3.3100	2.7114	3.8100	4.7508
2.3200	1.1809	2.8200	1.7051	3.3200	2.7398	3.8200	4.8083
2.3300	1.1882	2.8300	1.7196	3.3300	2.7685	3.8300	4.8666
2.3400	1.1956	2.8400	1.7344	3.3400	2.7976	3.8400	4.9257
2.3500	1.2031	2.8500	1.7494	3.3500	2.8272	3.8500	4.9857
2.3600	1.2107	2.8600	1.7646	3.3600	2.8571	3.8600	5.0466
2.3700	1.2184	2.8700	1.7799	3.3700	2.8875	3.8700	5.1084
2.3800	1.2262	2.8800	1.7955	3.3800	2.9183	3.8800	5.1711
2.3900	1.2341	2.8900	1.8113	3.3900	2.9495	3.8900	5.2348
2.4000	1.2422	2.9000	1.8274	3.4000	2.9812	3.9000	5.2993
2.4100	1.2503	2.9100	1.8436	3.4100	3.0133	3.9100	5.3648
2.4200	1.2586	2.9200	1.8600	3.4200	3.0459	3.9200	5.4313
2.4300	1.2670	2.9300	1.8767	3.4300	3.0789	3.9300	5.4988
2.4400	1.2756	2.9400	1.8936	3.4400	3.1124	3.9400	5.5673
2.4500	1.2842	2.9500	1.9108	3.4500	3.1463	3.9500	5.6368
2.4600	1.2930	2.9600	1.9281	3.4600	3.1807	3.9600	5.7073
2.4700	1.3019	2.9700	1.9457	3.4700	3.2156	3.9700	5.7789
2.4800	1.3109	2.9800	1.9636	3.4800	3.2510	3.9800	5.8515
2.4900	1.3201	2.9900	1.9817	3.4900	3.2869	3.9900	5.9252
2.5000	1.3293	3.0000	2.0000	3.5000	3.3233	4.0000	6.0000

Gamma Function

n	Γ(n)	n	Γ(n)	n	Γ(n)	n	Γ(n)
4.0100	6.0759	4.5100	11.7945	5.0100	24.3645	5.5100	53.1933
4.0200	6.1530	4.5200	11.9599	5.0200	24.7351	5.5200	54.0589
4.0300	6.2312	4.5300	12.1280	5.0300	25.1118	5.5300	54.9396
4.0400	6.3106	4.5400	12.2986	5.0400	25.4948	5.5400	55.8358
4.0500	6.3912	4.5500	12.4720	5.0500	25.8843	5.5500	56.7477
4.0600	6.4730	4.5600	12.6482	5.0600	26.2803	5.5600	57.6757
4.0700	6.5560	4.5700	12.8271	5.0700	26.6829	5.5700	58.6200
4.0800	6.6403	4.5800	13.0089	5.0800	27.0922	5.5800	59.5809
4.0900	6.7258	4.5900	13.1936	5.0900	27.5085	5.5900	60.5588
4.1000	6.8126	4.6000	13.3813	5.1000	27.9317	5.6000	61.5539
4.1100	6.9008	4.6100	13.5719	5.1100	28.3621	5.6100	62.5666
4.1200	6.9902	4.6200	13.7656	5.1200	28.7997	5.6200	63.5972
4.1300	7.0811	4.6300	13.9624	5.1300	29.2448	5.6300	64.6460
4.1400	7.1733	4.6400	14.1624	5.1400	29.6973	5.6400	65.7135
4.1500	7.2669	4.6500	14.3655	5.1500	30.1575	5.6500	66.7998
4.1600	7.3619	4.6600	14.5720	5.1600	30.6255	5.6600	67.9054
4.1700	7.4584	4.6700	14.7817	5.1700	31.1014	5.6700	69.0306
4.1800	7.5563	4.6800	14.9948	5.1800	31.5853	5.6800	70.1758
4.1900	7.6557	4.6900	15.2114	5.1900	32.0775	5.6900	71.3414
4.2000	7.7567	4.7000	15.4314	5.2000	32.5781	5.7000	72.5277
4.2100	7.8592	4.7100	15.6550	5.2100	33.0872	5.7100	73.7352
4.2200	7.9632	4.7200	15.8822	5.2200	33.6049	5.7200	74.9642
4.2300	8.0689	4.7300	16.1131	5.2300	34.1314	5.7300	76.2152
4.2400	8.1762	4.7400	16.3478	5.2400	34.6670	5.7400	77.4884
4.2500	8.2851	4.7500	16.5862	5.2500	35.2117	5.7500	78.7845
4.2600	8.3957	4.7600	16.8285	5.2600	35.7656	5.7600	80.1038
4.2700	8.5080	4.7700	17.0748	5.2700	36.3291	5.7700	81.4467
4.2800	8.6220	4.7800	17.3250	5.2800	36.9022	5.7800	82.8136
4.2900	8.7378	4.7900	17.5794	5.2900	37.4851	5.7900	84.2052
4.3000	8.8554	4.8000	17.8378	5.3000	38.0780	5.8000	85.6216
4.3100	8.9747	4.8100	18.1005	5.3100	38.6811	5.8100	87.0636
4.3200	9.0960	4.8200	18.3675	5.3200	39.2946	5.8200	88.5315
4.3300	9.2191	4.8300	18.6389	5.3300	39.9186	5.8300	90.0259
4.3400	9.3441	4.8400	18.9147	5.3400	40.5534	5.8400	91.5472
4.3500	9.4711	4.8500	19.1950	5.3500	41.1991	5.8500	93.0960
4.3600	9.6000	4.8600	19.4800	5.3600	41.8559	5.8600	94.6727
4.3700	9.7309	4.8700	19.7696	5.3700	42.5241	5.8700	96.2780
4.3800	9.8639	4.8800	20.0640	5.3800	43.2039	5.8800	97.9122
4.3900	9.9989	4.8900	20.3632	5.3900	43.8953	5.8900	99.5761
4.4000	10.1361	4.9000	20.6674	5.4000	44.5988	5.9000	101.2701
4.4100	10.2754	4.9100	20.9765	5.4100	45.3145	5.9100	102.9949
4.4200	10.4169	4.9200	21.2908	5.4200	46.0426	5.9200	104.7509
4.4300	10.5606	4.9300	21.6103	5.4300	46.7833	5.9300	106.5389
4.4400	10.7065	4.9400	21.9351	5.4400	47.5370	5.9400	108.3594
4.4500	10.8548	4.9500	22.2652	5.4500	48.3037	5.9500	110.2129
4.4600	11.0053	4.9600	22.6009	5.4600	49.0838	5.9600	112.1003
4.4700	11.1583	4.9700	22.9420	5.4700	49.8775	5.9700	114.0219
4.4800	11.3136	4.9800	23.2889	5.4800	50.6850	5.9800	115.9787
4.4900	11.4714	4.9900	23.6415	5.4900	51.5067	5.9900	117.9711
4.5000	11.6317	5.0000	24.0000	5.5000	52.3427	6.0000	120.0000

Gamma Function

n	$\Gamma(n)$	n	$\Gamma(n)$	n	$\Gamma(n)$	n	$\Gamma(n)$
6.0100	122.0661	6.5100	293.0953	7.0100	733.6171	7.5100	1908.0504
6.0200	124.1700	6.5200	298.4052	7.0200	747.5034	7.5200	1945.6019
6.0300	126.3123	6.5300	303.8161	7.0300	761.6632	7.5300	1983.9192
6.0400	128.4940	6.5400	309.3305	7.0400	776.1037	7.5400	2023.0216
6.0500	130.7156	6.5500	314.9500	7.0500	790.8292	7.5500	2062.9221
6.0600	132.9781	6.5600	320.6770	7.0600	805.8471	7.5600	2103.6414
6.0700	135.2820	6.5700	326.5134	7.0700	821.1620	7.5700	2145.1926
6.0800	137.6285	6.5800	332.4616	7.0800	836.7813	7.5800	2187.5977
6.0900	140.0181	6.5900	338.5236	7.0900	852.7099	7.5900	2230.8706
6.1000	142.4518	6.6000	344.7020	7.1000	868.9559	7.6000	2275.0332
6.1100	144.9303	6.6100	350.9986	7.1100	885.5239	7.6100	2320.1006
6.1200	147.4546	6.6200	357.4164	7.1200	902.4222	7.6200	2366.0967
6.1300	150.0255	6.6300	363.9571	7.1300	919.6564	7.6300	2413.0356
6.1400	152.6441	6.6400	370.6239	7.1400	937.2346	7.6400	2460.9426
6.1500	155.3111	6.6500	377.4185	7.1500	955.1622	7.6500	2509.8330
6.1600	158.0274	6.6600	384.3443	7.1600	973.4484	7.6600	2559.7332
6.1700	160.7941	6.6700	391.4035	7.1700	992.0996	7.6700	2610.6589
6.1800	163.6120	6.6800	398.5985	7.1800	1011.1224	7.6800	2662.6379
6.1900	166.4825	6.6900	405.9326	7.1900	1030.5265	7.6900	2715.6887
6.2000	169.4060	6.7000	413.4079	7.2000	1050.3174	7.7000	2769.8330
6.2100	172.3841	6.7100	421.0280	7.2100	1070.5054	7.7100	2825.0981
6.2200	175.4175	6.7200	428.7951	7.2200	1091.0967	7.7200	2881.5032
6.2300	178.5075	6.7300	436.7129	7.2300	1112.1016	7.7300	2939.0776
6.2400	181.6549	6.7400	444.7835	7.2400	1133.5264	7.7400	2997.8406
6.2500	184.8612	6.7500	453.0110	7.2500	1155.3823	7.7500	3057.8242
6.2600	188.1272	6.7600	461.3976	7.2600	1177.6760	7.7600	3119.0474
6.2700	191.4543	6.7700	469.9473	7.2700	1200.4185	7.7700	3181.5435
6.2800	194.8435	6.7800	478.6627	7.2800	1223.6171	7.7800	3245.3328
6.2900	198.2962	6.7900	487.5479	7.2900	1247.2832	7.7900	3310.4497
6.3000	201.8134	6.8000	496.6054	7.3000	1271.4244	7.8000	3376.9170
6.3100	205.3968	6.8100	505.8398	7.3100	1296.0537	7.8100	3444.7686
6.3200	209.0471	6.8200	515.2535	7.3200	1321.1777	7.8200	3514.0283
6.3300	212.7661	6.8300	524.8511	7.3300	1346.8097	7.8300	3584.7332
6.3400	216.5549	6.8400	534.6356	7.3400	1372.9580	7.8400	3656.9072
6.3500	220.4150	6.8500	544.6116	7.3500	1399.6354	7.8500	3730.5891
6.3600	224.3476	6.8600	554.7819	7.3600	1426.8507	7.8600	3805.8037
6.3700	228.3544	6.8700	565.1516	7.3700	1454.6176	7.8700	3882.5908
6.3800	232.4366	6.8800	575.7236	7.3800	1482.9451	7.8800	3960.9780
6.3900	236.5959	6.8900	586.5032	7.3900	1511.8477	7.8900	4041.0063
6.4000	240.8335	6.9000	597.4933	7.4000	1541.3342	7.9000	4122.7306
6.4100	245.1514	6.9100	608.6996	7.4100	1571.4203	7.9100	4206.1133
6.4200	249.5509	6.9200	620.1256	7.4200	1602.1152	7.9200	4291.2651
6.4300	254.0334	6.9300	631.7754	7.4300	1633.4349	7.9300	4378.2036
6.4400	258.6011	6.9400	643.6547	7.4400	1665.3906	7.9400	4466.9629
6.4500	263.2550	6.9500	655.7667	7.4500	1697.9950	7.9500	4557.5786
6.4600	267.9975	6.9600	668.1176	7.4600	1731.2637	7.9600	4650.0986
6.4700	272.8297	6.9700	680.7109	7.4700	1765.2081	7.9700	4744.5557
6.4800	277.7539	6.9800	693.5528	7.4800	1799.8456	7.9800	4840.9985
6.4900	282.7716	6.9900	706.6470	7.4900	1835.1874	7.9900	4939.4629
6.5000	287.8849	7.0000	720.0000	7.5000	1871.2517	8.0000	5040.0000

Gamma Function

n	Γ(n)	n	Γ(n)	n	Γ(n)	n	Γ(n)
8.0100	5142.6606	8.5100	14329.4697	9.0100	41192.7070	9.5100	121943.7969
8.0200	5247.4688	8.5200	14630.9121	9.0200	42084.6953	9.5200	124655.3359
8.0300	5354.4927	8.5300	14938.9092	9.0300	42996.5742	9.5300	127428.8906
8.0400	5463.7700	8.5400	15253.5820	9.0400	43928.7188	9.5400	130265.6094
8.0500	5575.3521	8.5500	15575.0781	9.0500	44881.5820	9.5500	133166.9062
8.0600	5689.2749	8.5600	15903.5146	9.0600	45855.5547	9.5600	136134.0625
8.0700	5805.6143	8.5700	16239.1074	9.0700	46851.3047	9.5700	139169.1562
8.0800	5924.4116	8.5800	16581.9902	9.0800	47869.2461	9.5800	142273.4688
8.0900	6045.7188	8.5900	16932.3242	9.0900	48909.8711	9.5900	145448.6562
8.1000	6169.5806	8.6000	17290.2344	9.1000	49973.5977	9.6000	148696.0156
8.1100	6296.0747	8.6100	17655.9668	9.1100	51061.1602	9.6100	152017.8594
8.1200	6425.2466	8.6200	18029.6543	9.1200	52173.0078	9.6200	155415.6406
8.1300	6557.1558	8.6300	18411.4805	9.1300	53309.6797	9.6300	158891.0625
8.1400	6691.8481	8.6400	18801.5801	9.1400	54471.6328	9.6400	162445.6406
8.1500	6829.4102	8.6500	19200.2207	9.1500	55659.6914	9.6500	166081.8906
8.1600	6969.8911	8.6600	19607.5547	9.1600	56874.3047	9.6600	169801.4062
8.1700	7113.3540	8.6700	20023.7734	9.1700	58116.1055	9.6700	173606.1250
8.1800	7259.8521	8.6800	20449.0371	9.1800	59385.5820	9.6800	177497.6250
8.1900	7409.4775	8.6900	20883.6270	9.1900	60683.6133	9.6900	181478.7188
8.2000	7562.2842	8.7000	21327.7109	9.2000	62010.7266	9.7000	185551.0938
8.2100	7718.3447	8.7100	21781.5059	9.2100	63367.6016	9.7100	189716.9219
8.2200	7877.7256	8.7200	22245.2266	9.2200	64754.9023	9.7200	193977.9688
8.2300	8040.4858	8.7300	22719.0449	9.2300	66173.1953	9.7300	198337.2812
8.2400	8206.7314	8.7400	23203.2871	9.2400	67623.4688	9.7400	202796.7500
8.2500	8376.5215	8.7500	23698.1367	9.2500	69106.3047	9.7500	207358.7031
8.2600	8549.9346	8.7600	24203.8340	9.2600	70622.4609	9.7600	212025.5625
8.2700	8727.0332	8.7700	24720.5664	9.2700	72172.5547	9.7700	216799.3438
8.2800	8907.9307	8.7800	25248.6895	9.2800	73757.6641	9.7800	221683.4844
8.2900	9092.6934	8.7900	25788.4023	9.2900	75378.4297	9.7900	226680.0938
8.3000	9281.4092	8.8000	26339.9766	9.3000	77035.6953	9.8000	231791.7969
8.3100	9474.1416	8.8100	26903.6133	9.3100	78730.1172	9.8100	237020.8281
8.3200	9671.0205	8.8200	27479.7012	9.3200	80462.8984	9.8200	242370.9844
8.3300	9872.1152	8.8300	28068.4609	9.3300	82234.7188	9.8300	247844.5156
8.3400	10077.5205	8.8400	28670.1797	9.3400	84046.5312	9.8400	253444.3906
8.3500	10287.3096	8.8500	29285.0918	9.3500	85899.0312	9.8500	259173.0625
8.3600	10501.6201	8.8600	29913.6172	9.3600	87793.5469	9.8600	265034.6250
8.3700	10720.5322	8.8700	30555.9902	9.3700	89730.8516	9.8700	271031.6250
8.3800	10944.1436	8.8800	31212.5391	9.3800	91711.9297	9.8800	277167.3125
8.3900	11172.5430	8.8900	31883.5117	9.3900	93737.6328	9.8900	283444.3438
8.4000	11405.8721	8.9000	32569.3535	9.4000	95809.3203	9.9000	289867.2188
8.4100	11644.2227	8.9100	33270.3555	9.4100	97927.9297	9.9100	296438.8438
8.4200	11887.7080	8.9200	33986.8477	9.4200	100094.4922	9.9200	303162.7500
8.4300	12136.4102	8.9300	34719.1172	9.4300	102309.9219	9.9300	310041.6562
8.4400	12390.4961	8.9400	35467.6445	9.4400	104575.7812	9.9400	317080.7500
8.4500	12650.0625	8.9500	36232.7461	9.4500	106893.0078	9.9500	324283.0938
8.4600	12915.2285	8.9600	37014.7852	9.4600	109262.8281	9.9600	331652.4688
8.4700	13186.1191	8.9700	37814.1406	9.4700	111686.1875	9.9700	339192.1875
8.4800	13462.8301	8.9800	38631.1328	9.4800	114164.8047	9.9800	346907.5625
8.4900	13745.5537	8.9900	39466.3086	9.4900	116699.7344	9.9900	354802.0625
8.5000	14034.3877	9.0000	40320.0000	9.5000	119292.2969	10.0000	362880.0000

附錄 H

卡方值表[*]

Critical Values of χ^2

Degrees of Freedom	$\chi^2_{0.995}$	$\chi^2_{0.990}$	$\chi^2_{0.975}$	$\chi^2_{0.950}$	$\chi^2_{0.900}$
1	0.0000393	0.0001571	0.0009821	0.0039321	0.0157908
2	0.0100251	0.0201007	0.0506356	0.102587	0.210720
3	0.0717212	0.114832	0.215795	0.351846	0.584375
4	0.206990	0.297110	0.484419	0.710721	1.063623
5	0.411740	0.554300	0.831211	1.145476	1.61031
6	0.675727	0.872085	1.237347	1.63539	2.20413
7	0.989265	1.239043	1.68987	2.16735	2.83311
8	1.344419	1.646482	2.17973	2.73264	3.48954
9	1.734926	2.087912	2.70039	3.32511	4.16816
10	2.15585	2.55821	3.24697	3.94030	4.86518
11	2.60321	3.05347	3.81575	4.57481	5.57779
12	3.07382	3.57056	4.40379	5.22603	6.30380
13	3.56503	4.10691	5.00874	5.89186	7.04150
14	4.07468	4.66043	5.62872	6.57063	7.78953
15	4.60094	5.22935	6.26214	7.26094	8.54675
16	5.14224	5.81221	6.90766	7.96164	9.31223
17	5.69724	6.40776	7.56418	8.67176	10.0852
18	6.26481	7.01491	8.23075	9.39046	10.8649
19	6.84398	7.63273	8.90655	10.1170	11.6509
20	7.43386	8.26040	9.59083	10.8508	12.4426
21	8.03366	8.89720	10.28293	11.5913	13.2396
22	8.64272	9.54249	10.9823	12.3380	14.0415
23	9.26042	10.19567	11.6885	13.0905	14.8479
24	9.88623	10.8564	12.4011	13.8484	15.6587
25	10.5197	11.5240	13.1197	14.6114	16.4734
26	11.1603	12.1981	13.8439	15.3791	17.2919
27	11.8076	12.8786	14.5733	16.1513	18.1138
28	12.4613	13.5648	15.3079	16.9279	18.9392
29	13.1211	14.2565	16.0471	17.7083	19.7677
30	13.7867	14.9535	16.7908	18.4926	20.5992
40	20.7065	22.1643	24.4331	26.5093	29.0505
50	27.9907	29.7067	32.3574	34.7642	37.6886
60	35.5346	37.4848	40.4817	43.1879	46.4589
70	43.2752	45.4418	48.7576	51.7393	55.3290
80	51.1720	53.5400	57.1532	60.3915	64.2778
90	59.1963	61.7541	65.6466	69.1260	73.2912
100	67.3276	70.0648	74.2219	77.9295	82.3581

Critical Values of χ^2 (Continued)

Degrees of Freedom	$\chi^2_{0.100}$	$\chi^2_{0.050}$	$\chi^2_{0.025}$	$\chi^2_{0.010}$	$\chi^2_{0.005}$
1	2.70554	3.84146	5.02389	6.63490	7.87944
2	4.60517	5.99147	7.37776	9.21034	10.5966
3	6.25139	7.81473	9.34840	11.3449	12.8381
4	7.77944	9.48773	11.1433	13.2767	14.8602
5	9.23635	11.0705	12.8325	15.0863	16.7496
6	10.6446	12.5916	14.4494	16.8119	18.5476
7	12.0170	14.0671	16.0128	18.4753	20.2777
8	13.3616	15.5073	17.5346	20.0902	21.9550
9	14.6837	16.9190	19.0228	21.6660	23.5893
10	15.9871	18.3070	20.4831	23.2093	25.1882
11	17.2750	19.6751	21.9200	24.7250	26.7569
12	18.5494	21.0261	23.3367	26.2170	28.2995
13	19.8119	22.3621	24.7356	27.6883	29.8194
14	21.0642	23.6848	26.1190	29.1413	31.3193
15	22.3072	24.9958	27.4884	30.5779	32.8013
16	23.5418	26.2962	28.8454	31.9999	34.2672
17	24.7690	27.5871	30.1910	33.4087	35.7185
18	25.9894	28.8693	31.5264	34.8053	37.1564
19	27.2036	30.1435	32.8523	36.1908	38.5822
20	28.4120	31.4104	34.1696	37.5662	39.9968
21	29.6151	32.6705	35.4789	38.9321	41.4010
22	30.8133	33.9244	36.7807	40.2894	42.7956
23	32.0069	35.1725	38.0757	41.6384	44.1813
24	33.1963	36.4151	39.3641	42.9798	45.5585
25	34.3816	37.6525	40.6465	44.3141	46.9278
26	35.5631	38.8852	41.9232	45.6417	48.2899
27	36.7412	40.1133	43.1944	46.9630	49.6449
28	37.9159	41.3372	44.4607	48.2782	50.9933
29	39.0875	42.5569	45.7222	49.5879	52.3356
30	40.2560	43.7729	46.9792	50.8922	53.6720
40	51.8050	55.7585	59.3417	63.6907	66.7659
50	63.1671	67.5048	71.4202	76.1539	79.4900
60	74.3970	79.0819	83.2976	88.3794	91.9517
70	85.5271	90.5312	95.0231	100.425	104.215
80	96.5782	101.879	106.629	112.329	116.321
90	107.565	113.145	118.136	124.116	128.299
100	118.498	124.342	129.561	135.807	140.169

附錄 I

t 分配值表

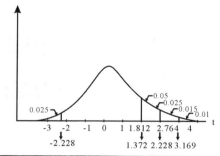

自 由 度	α				
	0.1	0.05	0.025	0.01	0.005
ν	$t_{0.1}$	$t_{0.05}$	$t_{0.025}$	$t_{0.01}$	$t_{0.005}$
1	3.078	6.314	12.706	31.821	63.657
2	1.886	2.920	4.303	6.965	9.925
3	1.638	2.353	3.182	4.541	5.841
4	1.533	2.132	2.776	3.747	4.604
5	1.476	2.015	2.571	3.365	4.032
6	1.440	1.943	2.447	3.143	3.707
7	1.415	1.895	2.365	2.998	3.499
8	1.397	1.860	2.306	2.896	3.355
9	1.383	1.833	2.262	2.821	3.250
10	1.372	1.812	2.228	2.764	3.169
11	1.363	1.796	2.201	2.718	3.106
12	1.356	1.782	2.179	2.681	3.055
13	1.350	1.771	2.160	2.650	3.012
14	1.345	1.761	2.145	2.624	2.977
15	1.341	1.753	2.131	2.602	2.947
16	1.337	1.746	2.120	2.583	2.921
17	1.333	1.740	2.110	2.567	2.898
18	1.330	1.734	2.101	2.552	2.878
19	1.328	1.729	2.093	2.539	2.861
20	1.325	1.725	2.086	2.528	2.845
21	1.323	1.721	2.080	2.518	2.831
22	1.321	1.717	2.074	2.508	2.819
23	1.319	1.714	2.069	2.500	2.807
24	1.318	1.711	2.064	2.492	2.797
25	1.316	1.708	2.060	2.485	2.787
26	1.315	1.706	2.056	2.479	2.779
27	1.314	1.703	2.052	2.473	2.771
28	1.313	1.701	2.048	2.467	2.763
29	1.311	1.699	2.045	2.462	2.756
30	1.310	1.697	2.042	2.457	2.750
40	1.303	1.684	2.021	2.423	2.704
60	1.296	1.671	2.000	2.390	2.660
120	1.290	1.661	1.984	2.358	2.626
∞	1.282	1.645	1.960	2.326	2.576

說明：1.　表中之數值為 t(t 分配的隨機變數)。

2.　t分配為常態分配之小樣本特例，分配曲線左右對稱。

3.　上圖之自由度 $\nu = 10$。

4.　表中 t 所對應之α，係指上圖中變數 t(橫軸)以右(較大)曲線所涵蓋的面積，即機率的和：$P\left(T > t_{\alpha,}(\nu)\right) = \alpha$

附錄 J

機率繪圖紙

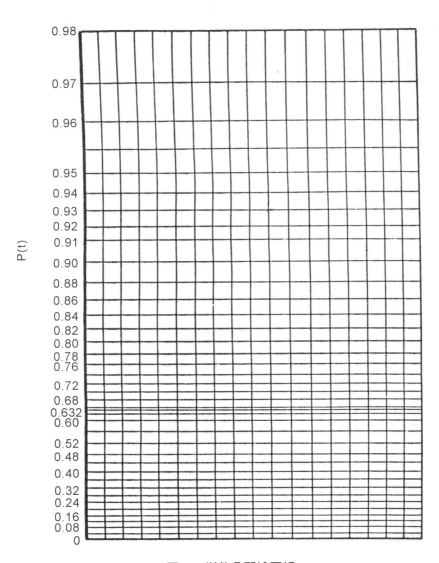

圖一　指數分配繪圖紙

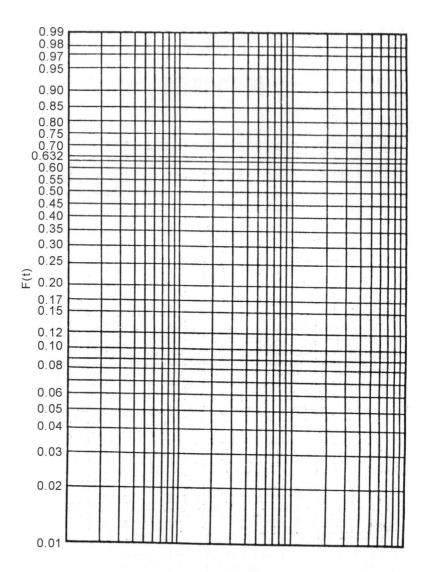

圖二　韋式分配繪圖紙

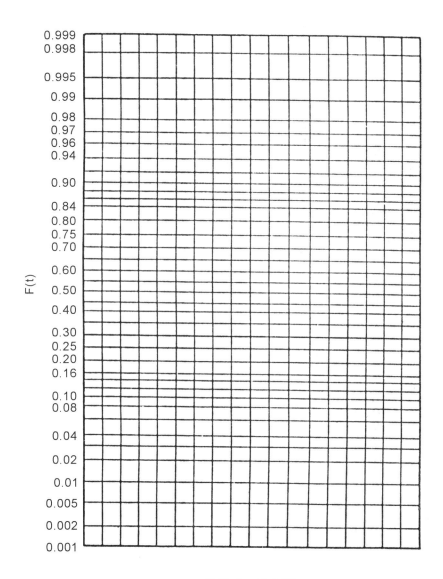

圖三　常態分配繪圖紙

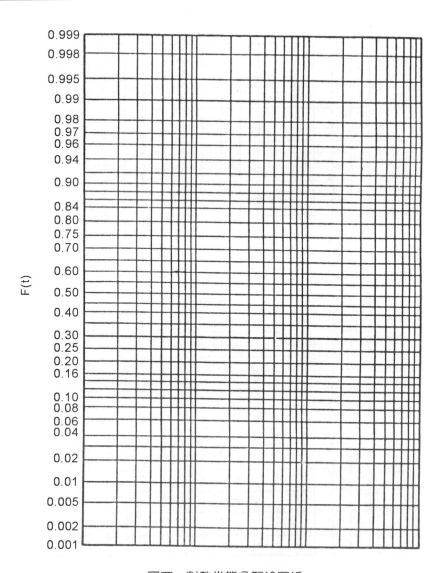

圖四　對數常態分配繪圖紙

參考文獻

1. Baxter, L. A. and Harche, F., "On the optimal assembly of series-parallel system", Operations Research Letters, 11, pp. 153-157, 1992.

2. David, R. and Alla, H., "Petri nets for modeling of dynamic systems-A survey", Automatica, **30**, (2), pp. 175-202, 1994.

3. Dummer, D. W. A., An Elementary Guide to Reliability, 3rd Edition, Pergamon Press, 1986.

4. Elsayed. A. Elsayed, Reliability Engineering, Addison Wesley Longman, Taipei, 1996.

5. Hura, G. S. and Atwood, J. W., "The use of Petri nets to analyze coherent fault trees", IEEE Transactions on Reliability, **37**, (5), pp. 469-474, 1988.

6. IEC 50(191), International Electrotechnical Vocabulary (IEV), Chapter 191—Dependability and quality of service, International Electrotechnical Commission, Geneva, 1990.

7. Kapur, K. C. and Lamberson, L. R., Reliability in Engineering Design, John Wiley, New York, 1977.

8. M. L. Shooman, Probabilistic Reliability: An Engineering Approach, McGraw-Hill, New York, 1968.

9. Patrick D. T. O'Connor, Practical Reliability Engineering, 4th Ed., John Wiley, Chichester, England, 2002.

10. Rausand, M. and Oien, K., "The basic concept of failure analysis", Reliability Engineering and System Safety, **53**, pp. 73-83, 1996.

11. W. G. Schneeweiss, Petri net for reliability modeling, Lilole-Verlag, Hagen, 1999.

12. S. S. Rao, Reliability-Based Design, McGraw-Hill, New York, 1992.

13. Yang, S. K. and Liu, T. S., 'Failure analysis for an airbag inflator by Petri nets', Quality and Reliability Engineering International, **13**, pp. 139-151, 1997.

14. Yang, S. K. and Liu, T. S., 'A Petri net approach to early failure detection and isolation for preventive maintenance', Quality and Reliability Engineering International, **14**(5), pp. 319-330, September 1998.

15. Yang, S. K. and Liu, T. S., 'State estimation for predictive maintenance using Kalman filter', Reliability Engineering and System Safety, **66**(1), pp. 29-39, October 1999.

16. Yang, S. K. and Liu, T. S., 'Implementation of Petri nets using a Field-Programmable Gate Array', Quality and Reliability Engineering International, **16**(2), pp. 99-116, March 2000.

17. Yang, S. K., 'An experiment of state estimation for predictive maintenance using Kalman filter on a DC motor', Reliability Engineering and System Safety, **75**(1), pp. 103-111, January 2002.

18. Yang, S. K., 'A System Reliability Evaluation Method By Finding Minimum Tie-sets Based On Petri Nets', Proceedings of The 2002 ROC Automatic Control Conference, TP125, Tainan, pp.1048-1055, March 2002。

19. Yang, S. K., 'A condition-based failure-prediction and processing-scheme for preventive maintenance', IEEE Transactions on Reliability, **52**(3), pp. 373-383, September 2003.

20. Yang, S. K., "A Petri-net approach to remote diagnosis for failures of cardiac pacemakers", Quality and Reliability Engineering International, **20**(7), pp. 761-776, November 2004。

21. 戴久永，可靠度導論，三民書局，台北，1990。

22. 楊善國，"發電廠中根據狀態而實施維護之研究"，中國機械工程學會第十五屆全國學術研討會論文集-製造與材料，台南，pp. 273-280, December 1998。

23. 楊善國，感測與量度工程，全華圖書公司，台北，March 2001 (4th Ed), ISBN 957-21-3184-2。